Cold Plasma in Materials Fabrication

AF794505

IEEE PRESS
445 Hoes Lane, PO Box 1331
Piscataway, NJ 08855-1331

1993 Editorial Board
William Perkins, *Editor in Chief*

R. S. Blicq
M. Eden
D. M. Etter
G. F. Hoffnagle
R. F. Hoyt
J. D. Irwin
S. Kartalopoulos
P. Laplante
A. J. Laub
M. Lightner
J. M. F. Moura
I. Peden
L. Shaw
D. J. Wells

Dudley R. Kay, *Director of Book Publishing*
Carrie Briggs, *Administrative Assistant*
Lisa Smey-Mizrahi, *Review Coordinator*

Karen Henley, *Production Editor*

Technical Reviewers for IEEE PRESS

Frank Barnes
University of Colorado

Hong Koo Kim
University of Pittsburgh

Cold Plasma in Materials Fabrication

From Fundamentals to Applications

Alfred Grill
IBM Research Division
T. J. Watson Research Center
Yorktown Heights, NY

IEEE PRESS

The Institute of Electrical and Electronics Engineers, Inc., New York

This book may be purchased at a discount from the publisher when ordered in bulk quantities. For more information contact:

IEEE PRESS Marketing
Attn: Special Sales
PO Box 1331
445 Hoes Lane
Piscataway, NJ 08855-1331
Fax: (908) 981-8062

© 1994 by the Institute of Electrical and Electronics Engineers, Inc.
345 East 47th Street, New York, NY 10017-2394

All rights reserved. No part of this book may be reproduced in any form, nor may it be stored in a retrieval system or transmitted in any form, without written permission from the publisher.

10 9 8 7 6 5 4 3 2 1

ISBN 0-7803-4714-5

IEEE Order Number PP4093

Library of Congress Cataloging-in-Publication Data

Grill, Alfred.
Cold plasma in materials technology: from fundamentals to applications / Alfred Grill.
p. cm.
Includes bibliographical references and index.
ISBN 0-7803-4714-5
1. Plasma engineering. 2. Low temperature plasmas. 3. Manufacturing processes. I. Title.
TA2020.G75 1994
621.044—dc20 93-42135
CIP

To Any,
Kalanit, Shirley, Yuval, and Yuval
with love.

Contents

Preface

Research and development for new, improved materials and their manufacturing processes have been rapidly accelerating during the last decades. This multi- and interdisciplinary effort encompasses many fields of science and technology such as physics, chemistry, and materials and chemical engineering. One of these fast-growing areas is processing by plasma assisted methods. A search of the published literature in the last 15 years shows an enormous growth in plasma assisted research and development activities, which reflects the significant task undertaken to understand these processes and use them in novel applications or as a useful substitute to established techniques.

Processing by plasma assisted techniques is being increasingly used in various areas of production and manufacturing as diverse as the automotive, aerospace, and biomedical industries and in the fabrication of microelectronic components. The current predominant trend in the microelectronics industry is to switch increasingly from wet to dry processing of materials by plasma assisted methods. The use of plasma techniques for deposition of films is expanding in response to requirements for processing at lower temperatures than those possible using chemical vapor deposition. As a result, a growing number of engineers and scientists from a wide range of disciplines are involved in plasma assisted processing, either in manufacturing or in research and development. To provide these engineers and scientists with an appropriate working background, a special course on plasma processing of materials must become part of their preparatory curriculum.

Although a number of books deal with cold, low-pressure plasmas and their applications, a textbook for a basic course appears to be needed, and this book is intended to fill that gap. The book will serve college seniors and graduate students in science (physics and chemistry) and engineering (materials, chemical, and electrical). It assumes the reader already has suitable knowledge of physics and chemistry. Two years of physics courses in a science or engineering program and a basic course in chemistry should provide the necessary basis for the comprehension of the material discussed in the book. The book is also designed to serve as an introductory and refreshment reference tool for the manufacturing engineer actively involved with plasma assisted techniques but also in need of detailed, in-depth knowledge of the field.

This book aims to present the fundamentals of physics and chemistry of cold plasmas, methods of generation, diagnostics, and state-of-the-art applications for processing of materials. The book is organized in eight chapters. The first chapter defines plasma in general as well as its main parameters and classifies the types of plasmas. All the chapters following focus on cold plasma only. The principles of the various methods to create and sustain a plasma are treated in the second chapter. The third chapter describes the reactions that can occur in cold plasmas and the interactions taking place between a plasma and a solid surface in contact with it. Reactors types, their design, and particular features are discussed in Chapter 4. Chapter 5 deals with plasma diagnostic methods, emphasizing quadrupole mass spectrometry, electrostatic Langmuir probes, and optical methods. Chapters 6–8 present a selection of applications of cold plasmas for processing solid surfaces: Chapter 6, devoted to surface modifications, covers surface activation, cleaning, ashing, oxidation, and hardening; Chapter 7 describes the deposition of coatings of inorganic films and polymerization; Chapter 8 centers on the use of plasma for dry etching. For the ease of the reader, a bibliographic list of relevant publications follows each chapter.

I wish to thank my wife Any for the critical reading of the manuscript and her constructive suggestions for improving its clarity.

Alfred Grill

List of Symbols

A	area
B	magnetic field
c	velocity of light
d	distance
d_s	thickness of plasma sheath
D	diffusion coefficient
e	electron charge
E	electric field
$f(W)$	energy distribution function
$f(v)$	velocity distribution function
f_{ce}	critical electron frequency
f_{ci}	critical ion frequency
h	Planck's constant
I	current
In	light intensity
J	current density
k	Boltzman's constant
L	characteristic reactor dimension
m, M	mass of particle
n	particle density, plasma density
N_D	number of particles in a Debye sphere
p	pressure
P	electrical power
q	electrical charge
Q	flow rate
r	radius
R	reaction rate
T	temperature
U, V	electrical potential, voltage
v	velocity
Vol	volume
W	energy
α	degree of ionization
ε_0	permittivity of free space
Γ	particle flux
λ_D	Debye length
λ	mean free path
μ	mobility
ν	collision frequency
ω	frequency
σ	cross section
τ_r	residence time

1

Fundamentals of Plasma

1.1 INTRODUCTION

Low-pressure plasma, cold plasma, nonequilibrium plasma, and glow discharge are some of the synonymously used terms to designate the same type of process. The technologies using these plasma assisted processes are generally referred to as plasma processing and include such diverse approaches as plasma assisted chemical vapor deposition (PACVD), plasma enhanced chemical vapor deposition (PECVD), ionitriding, and plasma etching. Plasma assisted processing is a critical technology used in the production of advanced microelectronics and in the production of present and future generations of large-scale integrated circuits. It would not be possible to manufacture very-large-scale integrated circuits (VLSI) computer chips without deposition of thin films by plasma assisted chemical vapor deposition or without plasma assisted etching, which enables etching of submicron-size features with vertical walls in silicon, metals, and dielectrics. Plasma processing also made possible the development of special materials with unique properties such as amorphous silicon or diamondlike carbon.

Plasma chemistry takes place under nonequilibrium conditions, and the reactions can occur while the gas or parts exposed to it remain at relatively low temperatures. The advantages of plasma processing are being exploited in various areas besides microelectronics. For example, the plasma assisted chemical vapor deposition technique called *ionitriding* allows replacement and upgrade of a conventional technology for surface hardening of metals, done by thermal nitriding, and thus achieves much more efficient surface hardening. The ionitriding technique enables control and adjustment of the properties of the hardened surface layers, not manageable while using the conventional method of nitriding.

At the base of the mentioned technologies is the cold plasma, a phenomenon similar to that occurring in fluorescent bulbs or neon lights, that is, an electrical discharge in a gas at low pressure. The phenomena occurring in cold plasmas are very complex and not yet fully understood. However, it is possible with the present knowledge of plasma physics and chemistry to adjust and control the composition of the gas mixtures and the parameters of the discharge to achieve required results in terms of processing and materials properties. The plasma assisted techniques allow increased production rates, precise production, and devising of materials with unique properties which evolve from the chemistry of cold plasmas.

The present book aims to provide a broad introduction to the physics and chemistry of cold plasmas, to present various diagnostic techniques used for studies, of plasma, for monitoring and optimization of plasma operations, and to discuss the state-of-the-art applications of cold plasmas.

1.2 DEFINITION OF PLASMA

Taking into consideration the energy of the particles constituting it, plasma is energetically the fourth state of the matter, apart from the solid, liquid, and gas states. Figure 1-1 presents schematically the ranges of temperature, or particle energy, in which each of the four forms of matter occur in nature. For the plasma state, the temperature range reflects only the energy of the heavy particles (not of the electrons) for reasons to be explained later.

Langmuir and collaborators were the first to study phenomena in plasma in the early 1920's while working on the development of vacuum tubes for large

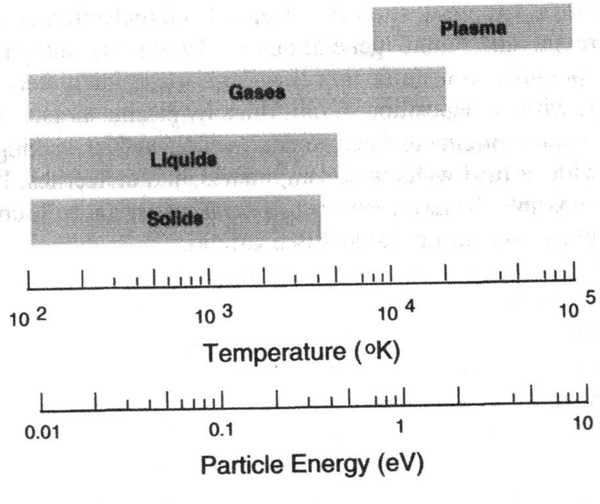

Fig. 1-1 State of matter versus temperature.

currents, and it was Langmuir who in 1929 used the term "plasma" for the first time [1] to describe ionized gases.

In a more rigorous way, a plasma can be defined as a *quasi-neutral gas* of charged and neutral particles characterized by a *collective behavior.*

Let us first define the collective property of the plasma. The behavior of a neutral gas is described by the kinetic theory of gases. According to this theory, in an ordinary neutral gas no forces act between the molecules of the gas (gravitational forces are considered negligible), and the particles travel in straight lines, with a distribution of velocities. The motion of the molecules is controlled by the collision among themselves and with the walls of the container. As a result of these collisions, the molecules of a neutral gas follow a random Brownian motion, as illustrated in Fig. 1-2 (a).

Assuming the particles of the neutral gas to be rigid spheres of radius r and their density n, the kinetic theory of the gases defines the *cross section for collision*, σ, and mean free path, λ, as

$$\sigma = \pi r^2 \tag{1.1}$$

$$\lambda = \frac{1}{\sigma n} \tag{1.2}$$

The average number of collisions per second, called the *collision frequency*, ν, and the *mean time between collisions*, τ, are given by

$$\nu = \frac{\bar{v}}{\lambda} \tag{1.3}$$

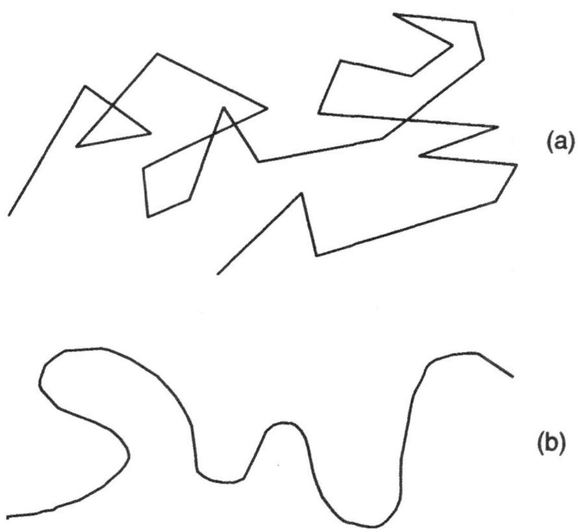

Fig. 1-2 Particle path in a neutral gas and under collective behavior in a plasma: (a) Brownian motion of a neutral gas molecule; (b) motion of a charged particle in a plasma.

$$\tau = \frac{1}{\nu} = \frac{\lambda}{\bar{v}} \tag{1.4}$$

where $\bar{v}$ is the *average velocity* of the molecules in the gas which is determined by its temperature, T:

$$\bar{v} = \left(\frac{kT}{M}\right)^{1/2} \tag{1.5}$$

M is the mass of the molecule, and k is the Boltzman constant. If the temperature of the gas is constant, the *collisional mean free path* is inversely proportional to the pressure in the system:

$$\lambda = \frac{ct}{p} \tag{1.6}$$

where ct = a constant depending on the gas
p = pressure in the gas

In a plasma, contrary to the preceding description, the motion of the particles can cause *local concentrations of positive and negative electric charges.* These charge concentrations create long-ranged Coulombic fields that affect the motion of charged particles far away from the charge concentrations. Thus elements of the plasma affect each other, even at large separations, giving the plasma its characteristic collective behavior. A charged particle in a plasma moves along a path which on average follows the electric field. Such a path is illustrated in Fig. 1-2(b). In some conditions, at low pressures, the effect of the long-range electromagnetic forces on the motion of the particles can be much stronger than the effect of the collisions between the particles. In such cases the plasma is called a *collisionless plasma.*

Local concentrations of charges in a plasma are confined to volumes of small dimensions of size λ_D, where λ_D is a characteristic dimension of the plasma, called the Debye length. This term will be explained in detail in Sec. 1.3.3. For a plasma to be stable, it is necessary that the dimensions of the system be much larger than the Debye length. Outside of these small volumes, of orders of tens of micrometers, the *charge density* of ions, which is generally approximately equal to the density of the ions, is equal to the density of electrons, making the plasma electrically neutral. As a result, the plasma is considered a quasi-neutral gas.

A plasma is usually obtained when sufficient energy, higher than the ionization energy, is added to atoms of a gas, causing ionization and production of ions and electrons. Parallel and concomitant to the ionization occurs the opposite process of recombination of electrons with ions to form neutral atoms or molecules. Although plasmas are commonly produced by electrical discharges in gases, a plasma can also be obtained when sufficient energy is provided to a liquid or a solid to cause its vaporization and ionization. Such plasma plumes can be obtained when high-density energy is supplied to a solid or a liquid by a laser. In a gas, a plasma is usually excited and sustained by providing to the gas electromagnetic energy in different forms: direct current, radio frequency, microwaves, and so on. Plasmas are often referred to as *gas discharges* because the most common way to produce plasma is by passing an electrical discharge through the gas.

1.3 PLASMA PARAMETERS

A plasma, especially one sustained in a mixture of molecular gases, contains a multitude of different neutral and charged particles. A group of identical particles in a plasma is commonly referred to as a *species*.

The plasma is broadly characterized by the following basic parameters:

- The *density of the neutral particles*, n_n.
- The *densities of the electrons and ions*, n_e and n_i. In the quasi-neutral state of plasma the densities of the electrons and of the ions are usually equal, $n_i = n_e = n$ and n is called the *plasma density*.
- The *energy distributions* of the neutral particles, $f_n(W)$; ions, $f_i(W)$; and electrons, $f_e(W)$.

The plasma density is an important parameter in plasma processing because the efficiency of the processes occurring in the plasma and their reaction rates are generally dependent directly on the density of the charged particles. As we shall see in Chapter 2, the electrons are the main factor responsible for the transfer of the energy from the external electric field to the discharge gas. Being electrically charged, both electrons and ions interact with the applied external electric field and are accelerated by absorbing energy from it. Because the electrons are the lightest particles in the plasma, they are easiest accelerated and absorb the largest amount of energy from the external field. The electrons then transfer through collisions energy to the molecules of the gas and cause their ionization and dissociation. The effectiveness of these processes increases with increasing electron density. Ions, too, play a significant role in the chemical reactions taking place in the plasma. Many of the reactions occurring in a plasma are controlled, or affected, by ion chemistry. It is therefore important to achieve high ion densities to increase the rates of reactions involving the ions.

As in any gaseous system, particles in the plasma are in continuous motion, inducing collisions between them. The collisions which take place between the particles in the plasma are of two types, elastic or inelastic. Collisions between electrons and heavy targets (i.e., neutral or charged particles) that do not result in an excitation of the target are called *elastic collisions*, whereas those collisions that leave the target in an excited state are called *inelastic collisions*.

The *energy transfer* W_{Tr} in an elastic collision between an electron and a heavy target is determined by the mass ratio of the particles

$$W_{Tr} = \frac{2m_e}{M} W \tag{1.7}$$

where M = mass of the heavy particle
W = energy of the electron
m_e = mass of electron

For an elastic collision of an electron with an argon atom, the fraction of transferred energy is therefore very small, about

$$\frac{W_{Tr}}{W} \approx \frac{1}{40{,}000} \tag{1.8}$$

On the other hand, a significant amount of energy is transferred in a collision between two electrons.

The electrons gain energy through acceleration by the electric field, which sustains the plasma and transfers that energy by inelastic collisions with the neutral gas molecules. The inelastic collisions between energetic electrons and the heavy species of the plasma result in excitation, ionization, or dissociation of the target if it is multiatomic. Energy transfer in an inelastic collision is not controlled by the mass ratio of the colliding particles. In an inelastic collision between two particles, the fraction of transferred energy is given by

$$\frac{W_{Tr}}{W} = \frac{M}{m_{in} + M} \tag{1.9}$$

where m_{in} is the mass of particle losing energy.

According to Eq. (1.9), in an inelastic collision between an electron and a heavy particle ($m_{in} = m_e \ll M$), the electron can transfer almost all its energy to the heavy particle, creating an energetic plasma species. The inelastic collisions therefore sustain the plasma by producing the particles that form it and giving the plasma its special features. Inelastic collisions involve energy transfer in amounts that vary from less than 0.1 eV (for rotational excitation of molecules) to more than 10 eV (for ionization).

Electron-electron collisions can also play a significant role in the energy transfer processes in the plasma. Their importance depends on the degree of ionization prevalent in the plasma. For degrees of ionization below 10^{-10}, the contribution of the electron-electron collisions to the energy transfer is negligible. However, in *electron cyclotron resonance* (*ECR*) plasmas, where the degree of ionization can be above 10^{-3}, electron-electron collisions dominate [2].

The relative contribution of each type of collision to the processes taking place in the plasma depends on additional plasma parameters, which will be discussed next and which derive from the previously described parameters.

1.3.1 The Degree of Ionization

The parameter that defines the density of the charged particles in the plasma is the *degree of ionization* of the gas. It specifies the fraction of the particles in the gaseous phase which are ionized. The degree of ionization, a, is defined as

$$\alpha = \frac{n_i}{n} \tag{1.10}$$

For plasmas sustained in low-pressure discharges, the degree of ionization is typically 10^{-6} to 10^{-3}. However, if the electrical discharge is assisted and confined by an additional magnetic field, the degree of ionization can reach values of 10^{-2} or higher, as for example, in an ECR plasma. Table 1-1 presents the range of values of the degree of ionization encountered in different low-pressure plasmas used for processing of solids.

TABLE 1-1 Ranges of Parameters for Various Low-Pressure Plasmas

Plasma Type	Pressure (torr)	Ion Density (cm^{-3})	Degree of Ionization
Deposition/etching	< 10	$< 10^{10}$	10^{-6}
Reactive ion Etching	10^{-2}–10^{-1}	10^{10}	10^{-6}–10^{-4}
Magnetron sputtering	10^{-3}	10^{11}	10^{-4}–10^{-2}
Electron cyclotron resonance	$< 10^{-4}$–10^{-2}	10^{12}	$< 10^{-1}$

The degree of ionization in a plasma is a function of the elements contained in the plasma. For example, in plasmas used in magnetron sputtering, the degree of ionization of the sputtered metal is higher than that of the process gas employed for the sputtering.

The value of the *critical ionization* is defined by [3]

$$a_c \approx 1.73 \times 10^{12} \sigma_{ea} T_e^2 \tag{1.11}$$

where σ_{ea} = electron-atom collision cross section at the average electron velocity, expressed in cm^2

T_e = electron temperature of the plasma, expressed in eV

The electron temperature will be defined later, in Sec. 1.3.2.1.

If the degree of ionization is much bigger than the critical ionization value, the charged particles behave as in a fully ionized gas.

1.3.2 Plasma Temperature

One of the physical parameters defining the state of a neutral gas in thermodynamic equilibrium is its temperature, which represents the *mean translational energy* of the molecules in the system. A plasma contains a mixture of particles with different electric charges and masses. At a first approximation, the plasma may be considered, thermally, as consisting of two systems: the first containing only electrons and the second containing the heavy species, that is, neutral atoms or molecules, ions, and neutral molecular fragments.

The electrons gain energy from the electric field, which energizes the plasma, and lose part of it by transfer to the second system through elastic or inelastic collisions. The system of heavy particles loses energy to the surroundings, either by radiation or by heat transfer to the walls of the vessel containing the plasma.

The electrons and the heavy species in the plasma can be considered approximately as two subsystems, each in its own thermal quasi-equilibrium.

The ions and electrons in the plasma can therefore be characterized by their specific different average temperatures: the *ion temperature*, T_i, and the *electron temperature*, T_e. Actually in some cases additional temperatures may characterize the particles in the plasma. For example, in the presence of a magnetic field, even a single plasma species, for example, the ions, is characterized by two different temperatures, one representing the translation of the ions parallel to the magnetic field, $T_{\|}$, and one representing the translation perpendicular to the magnetic field, $T_{\perp}$. This is caused by the fact that the forces acting on the species parallel to the magnetic field are different from those acting perpendicular to it [4].

The situation is even more complicated, as the heavy species in the plasma can be characterized by several temperatures at the same time, even in the absence of a magnetic field [5]: the *temperature of the gas*, T_g, which characterizes the translatory energy of the gas; the *excitation temperature*, T_{ex}, which characterizes the energy of the excited particles in the plasma; the *ionization temperature*, T_{ion}; the *dissociation temperature*, T_d, which characterize the energy of ionization and dissociation; and the *radiation temperature*, T_r, which characterizes the radiation energy. Thermodynamic equilibrium will exist in the plasma only if the following equation is satisfied:

$$T_g = T_{ex} = T_{ion} = T_d = T_r = T_e \tag{1.12}$$

Complete thermodynamic equilibrium cannot be achieved in the entire plasma because the radiation temperature, T_r, at the envelope of the plasma cannot equal the temperature in the plasma bulk. However, under certain laboratory conditions, it is possible to achieve local thermodynamic equilibrium in plasma in volumes of order of the mean free path length. If this happens, the plasma is called a *local thermodynamic equilibrium* (*LTE*) plasma. In low pressure plasmas, produced by direct current glow discharge or radio frequency excitation, the LTE conditions are generally not achieved. These plasmas are therefore called *non-LTE plasmas.*

In non-LTE plasmas the temperatures of the heavy particles are normally too small to promote chemical reactions in thermodynamic equilibrium. The electron temperature is therefore the most important temperature in non-LTE plasmas, among all those different temperatures mentioned previously. The fraction of electrons that will cause the different reactions in the plasma, the overall efficiency of the plasma processes, and the processing rates increase with increasing electron temperature. The electron temperature is discussed in further detail in the following section.

1.3.2.1 Electron Temperature

The *velocity distribution* function $f(v)$ for a system of particles is defined as the density of particles in the velocity space that satisfies the equation

$$n(cm^{-3}) = 4\pi \int_0^\infty f(v)v^2\, dv \tag{1.13}$$

where v = velocity
$f(v)$ = velocity distribution function (density in velocity space)
n = the density of the particles in the geometrical space

If it is assumed that the velocity distribution of the electrons in the plasma is isotropic, that the effects of inelastic collisions act only as a perturbation to the isotropy, and that the effects of the electric fields are negligible, then the velocity distribution is Maxwellian. The *Maxwellian distribution* assumes that the temperature of the electrons equals the temperature of the gas, $T_e = T_g$.

If the distribution of the electron velocities can be considered Maxwellian, then it can be described by [4, 6]

$$f(v) = n_e \left(\frac{m_e}{2\pi kT_e}\right)^{3/2} \exp\left(-\frac{m_e v^2}{2kT_e}\right) \tag{1.14}$$

The *electron energy distribution* function $f(W)$ is related to the velocity distribution function $f(v)$ through the relation [6]

$$f(W) = \frac{4\pi}{m_e}\, vf(v) \tag{1.15}$$

Therefore, the Maxwellian energy distribution function for the electrons is given by

$$f(W) = 2.07 W_{av}^{-3/2}\, W^{1/2} \exp\left(\frac{-1.5W}{W_{av}}\right) \tag{1.16}$$

where W_{av} is the *average energy of electrons.*
It can be shown that the average energy of the electrons is related to their temperature by

$$W_{av} = \frac{3}{2}\, kT_e \tag{1.17}$$

Due to the simplifying assumptions, the Maxwellian distribution provides only a first approximation of the electron energy (or velocity) distribution in the plasma.

The assumptions made for the Maxwellian distribution can be replaced in low-pressure plasmas by the following assumptions:

1. The electric field strength in the plasma is sufficiently low such that one can neglect the inelastic collisions, but large enough for the electron temperature to be much higher than the ion temperature, $T_e \gg T_i$.
2. The electric field is of sufficiently low frequency, that is, it is of a frequency ω much lower than the frequency of collisions ν.
3. The collision frequency is independent of the electron energy.

Under these assumptions the distribution of the electrons in the plasma is given by a *Druyvesteyn distribution* [6]. The Druyvesteyn distribution function gives a better approximation than the Maxwellian one for the electron energy distribution in the non-LTE plasmas.

The Druyvesteyn energy distribution of the electrons in the plasma is given by

$$f(W) = 1.04W_{av}^{-3/2}W^{1/2}\exp\left(-\frac{0.55W^2}{W_{av}^2}\right) \tag{1.18}$$

In the case of the Druyvesteyn distribution, the average electron energy and the electron temperature are functions of E_o/p, where E_o is the strength of the electric field and p is the pressure in the plasma. However, when the degree of ionization becomes large, the electron density also affects the energy distribution [6]. It should be emphasized that the Druyvesteyn distribution, as the Maxwellian one, gives only an approximation for a steady plasma. Numerical calculations have to be made to obtain a more accurate evaluation of the electron energy distribution.

Figure 1-3 illustrates Maxwellian and Druyvesteyn distributions for a sample of several average electron energies. As can be seen, the Druyvesteyn distribution is characterized by a shift toward higher electron energies, as compared to the Maxwellian one. As we shall see later, the reaction rates in the plasma are a function of interaction cross sections, which in turn depend on the energies of the particles. Some reactions in the plasma have an energetic threshold and will happen only if the energy of the participating electron is higher than the threshold value. An important fact illustrated by Fig. 1-3 is that both energy distributions are characterized by a *high-energy tail*. For an average electron energy of 5 eV, a significant amount of electrons have energies above 8 eV, reaching even values

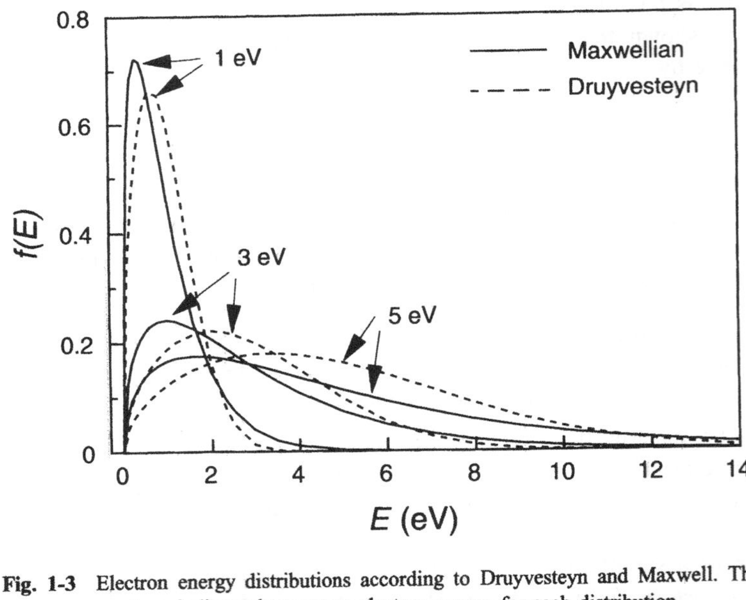

Fig. 1-3 Electron energy distributions according to Druyvesteyn and Maxwell. The numbers indicate the average electron energy for each distribution.

up to 14 eV. Thus the electrons in the high-energy tail of the distribution, though in small concentrations, have a significant impact on the overall reaction rates in the plasma. The Druyvesteyn distribution predicts a larger number of electrons to contribute to reactions requiring high energies.

As mentioned before, the different species in the plasma may be characterized by their distinct temperatures. The temperatures tend to equilibrate as the interaction between the two systems, that is, electrons and heavy particles, increases. This happens if either the pressure or the density of the electrons in the plasma increase, as shown in Fig. 1-4.

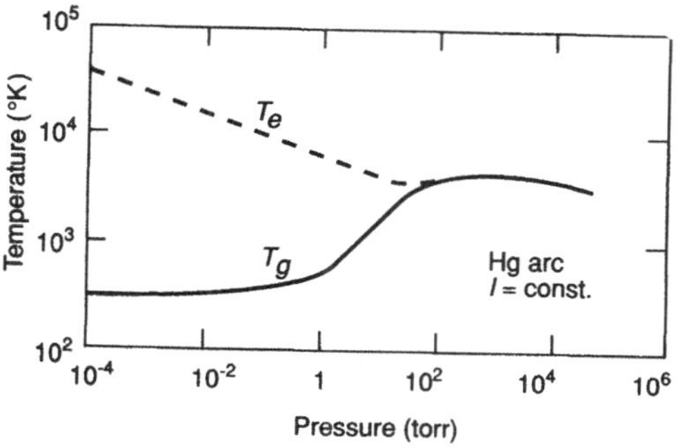

Fig. 1-4 Electron and ion temperatures as a function of pressure (from [7], reprinted with permission from R.F. Baddour and R.S. Timmins, *The Application of Plasmas to Chemical Pressing*, MIT Press, © 1967).

At low pressures, the electron temperature is much greater than the temperature of the gas, $T_e \gg T_g$. When the pressure in the plasma increases, the energy transfer from electron to neutrals increases, causing an increase in the temperature of the gas and decrease of the electron temperature (see Fig. 1-4). The electron and gas temperature converge to similar values at a pressure between 10 torr and 100 torr and the plasma becomes arclike. In arcs at atmospheric pressure the two temperatures are equal, $T_e = T_g$. When the two temperatures are about the same, the distribution of the species in the plasma can be described by equilibrium relations, while in the case when $T_e \gg T_g$, the distribution of active species is best represented by the electron temperature, T_e.

Although the temperatures of the electrons in the types of plasmas to be discussed in the following chapters are several times 10^4 °K (several electron volts), it does not imply that these plasmas are hot. Typical electron densities in these plasmas are about 10^{10} cm^{-3}, as compared to the density of particles in a gas at atmospheric pressure which is about $\sim 2.7 \times 10^{19}$ cm^{-3}. Due to the very low density and the very low heat capacity of the electrons, the amount of heat transferred by the electrons to the gas (heavy neutral and ionized particles) and to the walls of the container is very small. Thus, the term "cold plasma" derives its meaning from the small amount of heat transferred to the gas or solids in contact with it.

1.3.3 Debye Length

Another important parameter of a plasma is its *Debye length*. If an electric field is created in the plasma, the charged particles will react to reduce the effect of the field. The lighter, more mobile, electrons will respond fastest to reduce the electric field. If a plasma had an excess of positive or negative particles, this excess would create an electric field and the electrons will move to cancel the charge.

The response of charged particles to reduce the effect of local electric fields is called *Debye shielding* and the shielding gives the plasma its quasi-neutrality characteristic. Let assume that an electric potential is applied between two surfaces immersed in a plasma. The surfaces will attract equal amounts of charged particles of opposite sign. The concentration of charged particles near the two surfaces will shield the charged surfaces from the plasma bulk, which will remain neutral. The applied electrical potential will therefore develop mostly near the surfaces, over a distance λ_D, called the *Debye length* and defined by

$$\lambda_D = \left(\frac{\varepsilon_o k T_e}{n_e e^2}\right)^{1/2} \tag{1.19}$$

where ε_o = the permittivity of the free space
e = the charge of the electron

To clarify the meaning of the Debye length, let's assume that a positive electric charge, q, is inserted in a plasma that is initially quasi-neutral. The charge will create an electric potential, which in free space would be [2]

$$V_o = \frac{q}{4\pi\varepsilon_0 d} \tag{1.20}$$

where d is the distance from the charge. In the plasma, the potential is affected by the plasma electrons and ions and its value can be obtained by solving Poisson's equation,

$$\nabla^2 V = -\frac{\rho}{\varepsilon_0} \tag{1.21}$$

where ρ is the total charge density in the plasma and is given by:

$$\rho = e(n_i - n_e) + q\delta(d) \tag{1.22}$$

where $\delta(d)$ is the Dirac δ function, indicating that q is a point charge.

The potential V changes the electron density, and assuming that the electrons are in thermodynamic equilibrium at temperature T, the density can be calculated as [2]:

$$n_e = n \exp\left(\frac{eV}{kT}\right) \tag{1.23}$$

As it can be assumed that $eV \ll kT$, the Poisson's equation can be rewritten using Eq. (1.22) as

$$\nabla^2 V = \frac{V}{\lambda_D} + q\delta(d) \tag{1.24}$$

with λ_D defined by Eq. (1.19). The solution of Eq. (1.24) is

$$V(d) = \frac{q}{4\pi\varepsilon_0 d} \exp\left(-\frac{d}{\lambda_D}\right) = V_0 \exp\left(-\frac{d}{\lambda_D}\right) \tag{1.25}$$

Equation (1.25) shows that the plasma changes the potential of free space, V_0, causing its attenuation with a decay length equal to the Debye length, λ_D. This attenuation of the potential produced by a local charge in the plasma is the Debye shielding effect.

For estimation purposes and taking into account that $T(°\mathrm{K}) = 11{,}600\ T(\mathrm{eV})$, it is convenient to calculate the Debye length from

$$\lambda_D(\mathrm{cm}) = 6.93\left[\frac{T_e\,(°\mathrm{K})}{n_e\,(\mathrm{cm}^{-3})}\right]^{1/2} = 743\left[\frac{T_e\,(\mathrm{eV})}{n_e\,(\mathrm{cm}^{-3})}\right]^{1/2} \tag{1.26}$$

An example of typical values found in a cold plasma is

$$T_e = 1\,\mathrm{eV}, \quad n_e = 10^{10}\,\mathrm{cm}^{-3}, \quad \text{and } \lambda_D = 74\,\mu\mathrm{m}$$

As indicated by Eq. (1.19), the Debye length decreases with increasing electron density. An ionized gas is considered a plasma only if the density of the charged particles is large enough such that $\lambda_D \ll L$, where L is the dimension of the system. If this condition is satisfied, local concentrations of electric charges which may occur in the plasma are shielded out by the Debye shielding effect over distances smaller than the Debye length. Outside these volumes of charge concentrations the plasma bulk is quasi-neutral. The Debye length, λ_D, is therefore the characteristic dimension of regions in which breakdown of neutrality (formation of local concentrations of charges) can occur in a plasma.

Another plasma parameter related to the Debye length is the *number of particles*, N_D, in a *Debye sphere*, that is, in a sphere of radius equal to λ_D. The solution of Poisson's equation given by Eq. (1.25) can be obtained only by assuming that the shielding effect is produced by a large number of electrons, or in other words, the shielding effect can occur only if the Debye sphere contains a large number of electrons. Due to the exponential decay of the potential, it can be assumed that the shielding is caused by the electrons in the Debye sphere, whose number is given by

$$N_D = \frac{4\pi}{3} n_e \lambda_D^3 = \frac{1.38 \times 10^3 T_e^{3/2}\,(°\mathrm{K})}{n_e^{1/2}} = \frac{1.718 \times 10^9 T_e^{3/2}\,(\mathrm{eV})}{n_e^{1/2}} \tag{1.27}$$

N_D has to be therefore much larger than unity to fulfill the collective characteristic of the plasma. For electron temperatures $T_e > 1$ eV and densities $n_e < 10^{12}$ cm^{-3}, the condition $N_D \gg 1$ is easily satisfied. In the cold plasmas N_D ranges from about 10^4 to 10^7 electrons in a Debye sphere.

1.3.4 Plasma Sheath

We shall proceed with a more detailed examination of what happens at a surface in contact with a plasma. Ions and electrons reaching the solid surface recombine and are lost from the plasma system. Electrons that have much higher thermal velocities than ions reach the surface faster and leave the plasma with a positive charge in the vicinity of the surface. An electric field that retards the electrons and accelerates the ions develops near the surface in such a way as to make the net current zero. As a result, the surface achieves a negative potential, relative to the plasma, or in other words, the surface is at a negative *self-bias* relative to the plasma.

The plasma is therefore always at a positive potential relative to any surface in contact with it. Because of the Debye shielding effect, the potential developed between the surface and the plasma bulk is confined to a layer of thickness of several Debye lengths. This layer of positive space charge that exists around all surfaces in contact with the plasma is called the *plasma sheath*.

The *sheath potential*, V_s, is the electrical potential developed across the plasma sheath. Only electrons having sufficiently high thermal energy will penetrate through the sheath and reach the surface, which, being negative relative to the plasma, tends to repel the electrons. The value of the sheath potential adjusts itself in such a way that the flux of these electrons is equal to the flux of ions reaching the surface. Its value is given for a planar surface by [8]

$$V_s = \frac{kT_e}{2e} \ln\left(\frac{m_e}{2.3m_i}\right) \tag{1.28a}$$

where V_s = sheath potential
m_i = mass of ion

For a spherical surface the expression changes to [9]

$$V_s = \frac{kT_e}{2e} \ln\left(\frac{\pi m_e}{2m_i}\right) \tag{1.28b}$$

Figure 1-5 shows a schematic diagram of a plasma sheath. As indicated in the figure, the plasma sheath is a region of positive *space charge*, almost devoid of negative charges. The *thickness of the plasma sheath*, d_s, is defined as the thickness of the region where the electron density is negligible and where the potential drop V_s occurs. As previously explained, the thickness of the plasma sheath is related to the Debye length. It also depends on the collisional mean free path in the plasma and is affected by external biases applied to the surface.

At higher pressures, when the collisional mean free path is of the same order of magnitude as the thickness of the plasma sheath, the latter can be estimated from [10]

$$d_s \approx \eta^{2/3} \times \lambda_D \tag{1.29a}$$

with

$$\eta = \frac{e(V_p - V_B)}{kT_e} \tag{1.29b}$$

where d_s = thickness of plasma sheath
V_B = the bias on the considered surface (self- or external bias)
V_p = the plasma potential, which will be defined in Sec. 5.2

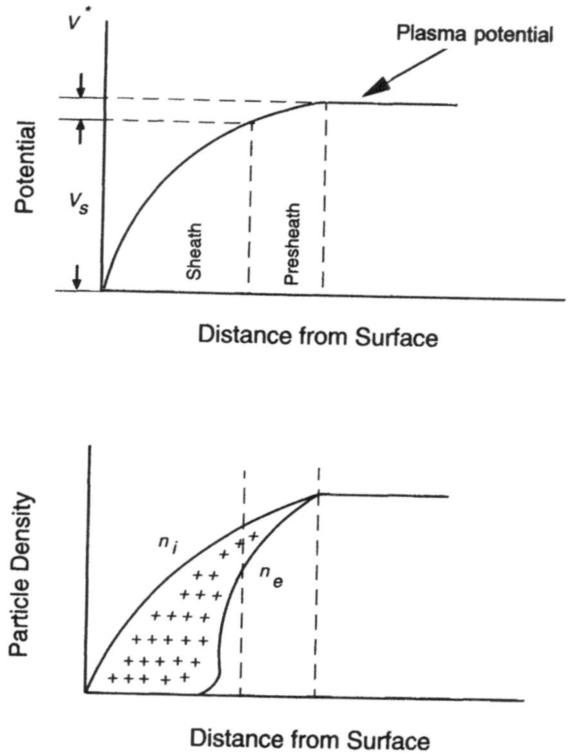

Fig. 1-5 Diagram of a plasma sheath.

At lower pressures, when the mean free path is much greater than the thickness of the plasma sheath, the thickness of the plasma sheath can be calculated from

$$d_s \approx 1.1\,\eta^{3/4} \times \lambda_D \tag{1.29c}$$

If the surface bias is small, of the order of magnitude of T_e, the sheath will only be a few Debye lengths thick. At pressures below ~ 0.05 torr the self-bias can reach values of tens to hundreds of volts, and the sheath thickness increases. If a bias of -100 V is applied to a surface in contact with an argon plasma in which $T_e \approx 1$ eV, the plasma sheath thickness is about $32\,\lambda_D$ [10].

It was experimentally found that the thickness of the plasma sheath is affected by more parameters than these figuring in Eq. 1.29. The relation between the thickness of the plasma sheath and those additional parameters is not clearly understood.

The thickness of the plasma sheath was found to be also dependent on the frequency of the electromagnetic field and the pressure in the system. For example, it was found in a certain system [11] that at 7 MHz, the plasma sheath thickness decreases from 12 mm to 7 mm when the pressure increases from 4 to

20 Pa, but at 27 MHz the plasma sheath thickness decreases from 6 mm to 4 mm when the pressure increases from 2.2 to 10 Pa. In the same system it was found that at some pressures the thickness appeared to be independent of the sheath voltage, while at other pressures it was found to be dependent on it [11].

The *ion current density* through the plasma sheath, J_i, is given by the *Child-Langmuir law* [4]:

$$J_i = 27.3\left(\frac{40}{m_i}\right)^{1/2} \frac{V_s^{3/2}}{d_s^2} \tag{1.30}$$

Equation (1.30) calculates the current density in units of $\mathrm{mA/cm^2}$ for V_s in kV and d_s in mm.

Between the sheath and the plasma bulk extends a quasi-neutral region called *presheath*. In that region there is a potential drop V^*, whose magnitude is of the order of $V^* = 0.5\,(kT_e/e)$ [4, 12].

The transport of ions and electrons to the surface through the sheath is done by ambipolar diffusion, discussed at length in Sec. 1.5. The flow of ions through the plasma sheath is controlled by the *Bohm sheath criterion*. This criterion establishes that only ions that enter the sheath with a directional velocity, v_i, greater than a critical value, v_c, will reach the substrate:

$$v_i > v_c \tag{1.31}$$

The value of the critical ion velocity can be calculated from [8]

$$v_c = \left(\frac{kT_e}{m_i}\right)^{1/2} \tag{1.32}$$

where v_c is the thermal velocity of the ions.

Equation (1.32) shows that according to the Bohm criterion, the minimum velocity required for an ion to reach the surface exposed to the plasma is a function of the electron temperature. This expresses the relationship which exists between the movements of ions and electrons in the plasma. The ions achieve the velocity v_i required by the the Bohm criterion by acceleration in the quasineutral presheath region.

If the conditions of the plasma are such that collisional mean free path, λ, is much larger than the Debye length ($\lambda \gg \lambda_D$), the plasma has a *collision-free sheath*.

1.3.5 Plasma Frequency

Although the plasma bulk is quasi-neutral, local perturbations from neutrality can occur in volumes smaller than the Debye sphere. Due to their low mass, the electrons will respond faster than the ions to the electric forces generated by the perturbation from neutrality. The response to the perturbation will be through oscillations. The frequency of these electron oscillations is called the *plasma*, or *Langmuir frequency*, ω_p, and is given by the relation [8]

$$\omega_p = \left(\frac{n_e e^2}{m_e \varepsilon_o}\right)^{1/2} = 18{,}000\ \pi n_e^{1/2}\ \mathrm{Hz} \tag{1.33}$$

for n_e expressed in cm^{-3}. For a typical plasma density of 10^{10} cm^{-3}, the plasma frequency is 9.10^8 Hz, a frequency much higher than the 13.56 MHz generally used to sustain a radio frequency (RF) discharge.

It can be deduced from Eq. (1.19) and Eq. (1.33) that the plasma frequency, ω_p, is related to the Debye length by the relation

$$\lambda_D \omega_p = \left(\frac{kT_e}{m_e}\right)^{1/2} \approx \bar{v}_e \tag{1.34}$$

Here, $\bar{v}_e$ is the *average thermal velocity* of the electrons in the plasma.

Equation (1.34) shows that the electrons can move over a distance of one Debye length during a period of the plasma oscillation. This indicates that if a perturbation of frequency $\omega < \omega_p$ is occurring in the plasma, the electrons can respond sufficiently fast to maintain the neutrality of the plasma. The oscillation frequency of the ions being much smaller due to their larger mass, only the electrons are able to maintain their own thermal equilibrium in the plasma. Plasma perturbations of frequencies higher than ω_p will not be shielded out through the response of the electrons.

1.4 CONDITIONS FOR PLASMA EXISTENCE

In Sec. 1.3.3 it was concluded that to sustain a plasma in a gas, two conditions have to be met:

$$\lambda_D \ll L \tag{1.35}$$

$$N_D \gg 1 \tag{1.36}$$

However, an additional third condition has to be fulfilled by a gas to become plasma. This condition is related to the frequency of collisions in the plasma. If the charged particles collide too frequently with neutral atoms, their motion is controlled by ordinary hydrodynamic forces rather than by electromagnetic forces. Under these circumstances the collective behavior condition is not satisfied, and the gas is not behaving as a plasma. If τ is the mean time between collisions of charged particles with neutral atoms, the product $\omega\tau$ has to be bigger than 1 for the gas to behave like a plasma rather than a neutral gas.

The three conditions that have to be satisfied by a plasma are therefore given by Eq. (1.35), Eq. (1.36), and Eq. (1.37):

$$\omega\tau > 1 \tag{1.37}$$

1.5 DIFFUSION OF CHARGED PARTICLES IN PLASMA

It was so far implicitly assumed that the plasmas are homogeneous. However, any plasma has a density gradient and has to be considered as having a nonuniform distribution of ions and electrons in a dense background of neutrals. Because of the concentration gradients, the plasma particles will tend to move by diffusion toward regions of lower density. As the plasma spreads out as a result of concentration gradient and electric field forces, the individual charged particles diffuse by undergoing a random walk and collide frequently with the neutral atoms. The electrical conductivity of the plasma caused by the movement of the charged particles is therefore controlled by the diffusion of the charged particles through it.

At very low charge concentrations, like those existing near the breakdown in a direct current glow discharge (described in Sec. 2.1), where the Debye length is of the same order of magnitude as the diffusion distance or system dimension, L, the electrons and ions will diffuse independently, and their flux will be controlled by their individual *diffusivities*, D_e and D_i, respectively. However, this is no longer true when the density of the charged particles increases above $n_e \approx n_i \geq 10^8$ cm^{-3} and the Debye length becomes much smaller than the dimensions of the system.

The *diffusion*, or *drift velocity*, v, of charged particles in an electric field is proportional to the field strength, E, and the proportionality factor is called the *mobility*, μ, of the particles:

$$\mu = \frac{v}{E} \tag{1.38}$$

The mobility of a charged particle is its drift velocity in an electric field of unity. Because of the much smaller mass of the electrons as compared to that of the ions, their mobility is much higher than the mobility of the ions in the plasma. The mobilities, μ, of the charged particles are correlated to their diffusion coefficients, D, by the Einstein equation

$$\mu = \frac{|q|D}{kT} \tag{1.39}$$

where q = the electric charge of the considered particles.

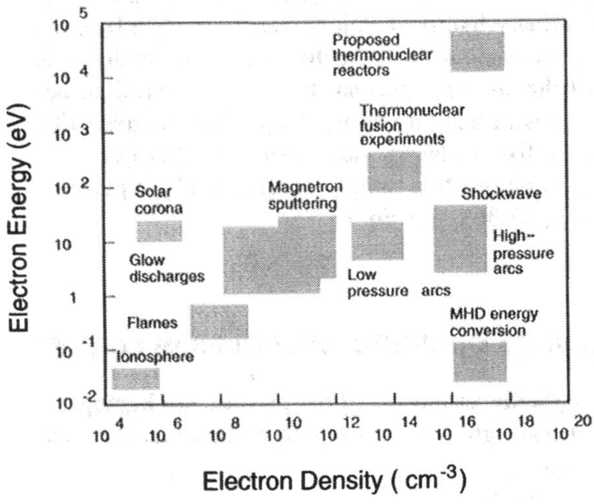

Fig. 1-6 Plasma types by electron density and temperature.

Due to the proportionality between the mobility and diffusion coefficient, as indicated by Eq. (1.39), the diffusion coefficient of the electrons is much higher than that of the ions. As a result, the electrons tend to diffuse toward regions of lower concentration much faster than the ions. This faster diffusion causes the formation of a space charge, which according to the Debye shielding effect (Sec. 1.3.3.) has to be contained over distances smaller than λ_D. A corresponding *space charge field*, E_{sc}, is formed.

The motion of the electrons is restrained by the space charge field, created by their tendency for faster diffusion. At the same time, the field that restrains the electrons causes the ions to diffuse faster than they would otherwise in the absence of the electrons. Consequently, both species of charged particles diffuse with the same velocity. Because it is also assumed that $n_e = n_i = n$, the *flux of the electrons*, Γ_e, is equal to the *flux of the ions*, Γ_i:

$$\Gamma_e = \Gamma_i = \Gamma \tag{1.40}$$

The described behavior is called *ambipolar diffusion* [13] because particles of opposite electric charges diffuse together due to their interaction.

The flux of charged particles is composed of two terms:

1. Flux associated with the motion induced by diffusion, caused by concentration gradients, ∇n; this term, $-D\nabla n$, is not affected by the electric charge of the particles.
2. Flux associated with the drift of the charged particles under the influence of the electric field; the sign of this term, $(\pm) n\mu E_{sc}$, depends on the charge of the particles.

The fluxes of the diffusing particles will therefore be given for electrons and ions, respectively, by

$$\Gamma_e = -D_e \nabla n_e - n_e \mu_e E_{sc} \tag{1.41}$$

$$\Gamma_i = -D_i \nabla n_i + n_i \mu_i E_{sc} \tag{1.42}$$

Assuming $n_e = n_i = n$ and $\Delta n_e = \Delta n_i = \Delta n$, Eq. (1.41) and Eq. (1.42) can be solved to give the value of the space charge field and the particle flux:

$$E_{sc} = -\left(\frac{D_e - D_i}{\mu_e + \mu_i}\right)\frac{\nabla n}{n} \tag{1.43}$$

$$\Gamma = -\left(\frac{D_i \mu_e + D_e \mu_i}{\mu_e + \mu_i}\right)\nabla n = -D_a \nabla n \tag{1.44}$$

The diffusion of the charged particle in the plasma is therefore not controlled by the individual diffusion coefficients of the electrons and ions but by the *ambipolar diffusion coefficient*, D_a:

$$D_a = \frac{D_e \mu_i + D_i \mu_e}{\mu_e + \mu_i} \tag{1.45}$$

As said before, the coefficient D_a reflects the interaction between plasma species of opposite electric charges.

Because $\mu_e \gg \mu_i$, Eq. (1.45) for the coefficient of ambipolar diffusion can be changed to

$$D_a = D_i\left(1 + \frac{D_e\mu_i}{D_i\mu_e}\right) \tag{1.46}$$

and using Eq. (1.39),

$$D_a = D_i\left(1 + \frac{T_e}{T_i}\right) \tag{1.47}$$

1.6 PLASMA TYPES

The plasma state exists in natural form in the cosmos or is created under unique conditions for specific purposes. The plasmas found in nature cover a very large range of electron densities and temperatures. As shown in Fig. 1-6, the plasma density, n_e, spans the range between 1 and 10^{20} cm^{-3}, while the electron temperature, T_e, can vary between 10^{-2} and 10^5 eV.

Solar winds are a continuous stream of charged particles with $n_e = 5$ cm^{-3} and $T_e = 50$ eV. The interstellar material contains a hydrogen plasma with a density of 1 cm^{-3}. The ionosphere, which extends approximately from 50 km upward from the earth's surface, is populated by a weak plasma with a density varying up to 10^6 cm^{-3} and an electron temperature of 0.1 eV, while the sun and stars have surface temperatures ranging from 5000 to more than 70,000 °K (0.5 to 7 eV). They consist entirely of plasma, the outer layer being partially ionized and the interior hot enough to be completely ionized. The temperature at the center of the sun is at about 2 keV.

In the quest for controlled thermonuclear fusion, it is necessary to create plasmas with electron temperatures above 10 keV and with ion densities of $(1 - 2) \times 10^{14}$ cm^{-3}. These values are required in order to obtain the nuclear fusion reaction between deuterium and tritium atoms, because reasonable cross-sections for the fusion reactions are obtained only for energies above 5 KeV. Central ions temperature of $T_i \approx 35$ KeV and electron temperature of $T_e = 15$ KeV have been reached in Tokomak fusion reactors [14].

Taking into account the wide ranges of parameters, the plasmas are classified into several categories:

- Plasmas in *complete thermodynamic equilibrium – CTE plasmas*. In a CTE plasma all temperatures discussed previously in Sec. 1.3.2 are equal. CTE plasmas exist only in stars or during the short interval of a strong explosion. They have no practical importance because they do not exist in controlled laboratory conditions.
- Plasmas in *local thermodynamic equilibrium – LTE plasmas*. These are plasmas in which all temperatures, except the radiation temperature, T_r, are

equal in each small volume of the plasma. The LTE plasmas are discussed in Sec. 1.6.1.

- Plasmas that are not in any *local thermodynamic equilibrium – non-LTE plasmas*. These plasmas, also named *cold plasmas*, are the subject of discussion in the following chapters.

The plasmas produced for research or manufacturing purposes are either LTE or non-LTE type plasmas, designated in daily use, respectively, as thermal and cold plasmas.

1.6.1 Thermal Plasmas

LTE plasmas can exist under two circumstances:

- When the heavy particles are very energetic, at temperatures of the order of 10^6–10^8 °K (10^2–10^4 eV)
- When the pressure is atmospheric, even at temperatures as low as 6000 °K

An increase of pressure in the plasma causes an increase in the number of collisions between the electrons and the heavy species. As a result, when the pressure in the system increases toward atmospheric pressure, the two systems tend to reach the same thermodynamic equilibrium, as shown in Fig. 1-4. For example, in electric arcs, or in plasma jets operating at pressures of about 1 atm, the temperature of the electrons is approximately equal to that of the gas, $T_e = T_g$. The temperature of the gas in the center of these plasmas can reach values of 20,000 to 30,000 °K. The high-intensity arc of plasma jets in inert gas atmosphere is used as a heating torch that is capable of delivering considerably higher temperatures and rates of heat transfer as compared to conventional torches.

Such LTE-plasmas at atmospheric pressure are called *thermal plasmas*. Their production and properties make them appropriate for deposition of coatings by plasma spraying processes and in extractive metallurgy, for reduction, or for smelting of ores.

The plasmas produced to create controlled thermonuclear fusion are LTE plasmas with very energetic heavy particles. The high energies can be obtained at low pressures of 10^{-8}–10^{-3} torr, and the main problem in heating these plasmas is to prevent the interaction of the energetic particles with the walls of the reactor. The interaction with the walls causes both loss of energy from the plasma and its contamination with particles sputtered from the walls of the plasma container. Continuous heating of the plasmas to the required temperatures of above 10 KeV (above 10^8 °K) and the confinement from interactions with the containing vessel are problems yet to be solved.

1.6.2 Cold Plasmas

As mentioned before, in low-pressure discharges, thermodynamic equilibrium is not reached, even at a local scale, between the electrons and the heavy particles, and these plasmas are of the non-LTE type. In the non-LTE plasmas the temperature of the electrons is much higher than that of the heavy particles and $T_e \gg T_i > T_g > T_{ex}$. The electrons can reach temperatures of 10^4–10^5 °K (1–10 eV), while the temperature of the gas, T_g, can be as low as room temperature. Therefore, such plasmas are called *cold plasmas*.

The cold plasmas have been developed specifically and purposefully based on their nonequilibrium properties and their capability to cause physical and chemical reactions with the gas at relatively low temperatures. Applications of cold plasmas are widespread and put to use in a variety of fields, from microelectronic fabrication to surface hardening of metals. In the following chapters we will concentrate on and examine in depth only the cold plasmas.

1.7 QUESTIONS

1. The value of the constant in Eq. (1.6) is $ct = 18.0 \times 10^{-3}$ cm.mbar for helium and $ct = 6.5 \times 10^{-3}$ cm.mbar for oxygen. Plot the mean free path versus pressure for both gases in the pressure range 10^{-5}–10 torr.
2. Plot the collision frequencies for the same conditions as in question 1, assuming that the gases are at room temperature.
3. The electron temperature in an argon plasma is 20,000 °K. Plot the energy distribution of the electrons assuming that it is a Druyvesteyn distribution.
4. a. Calculate the average energy of the electrons for the conditions stated in question 3.
   b. Calculate the fraction of electrons having energies higher than twice the average energy at those conditions.
   c. How will the values calculated in (a) and (b) change in an oxygen plasma at identical conditions?
5. a. Calculate the Debye length in a helium plasma having a density of 5.10^9 cm^{-3} and an average electron energy of 4 eV.
   b. How will the Debye length change if the plasma density increases to 10^{11} cm^{-3} and the electron temperature remains the same?
   c. What is the number of particles in a Debye sphere in each of the previous cases?
6. a. Calculate the sheath potential at a planar surface in contact with an argon plasma sustained at a pressure of 1 torr and having an electron temperature of 1 eV.
   b. What is the thickness of the plasma sheath if the plasma potential is 20 V ?
   c. Calculate the thickness of the plasma sheath in a plasma sustained at 10^{-4} torr, assuming the same electron temperature and plasma potential as before.
7. What are the ion current densities in argon plasmas sustained at the conditions corresponding to questions 6 (b) and (c)?

8. What is the lowest density of a plasma that could be sustained in a reactor whose smallest dimension is 5 cm, at an electron temperature of 1 eV? How would that density change for an electron temperature of 10 eV?
9. In terms of thermodynamic equilibrium, what type of plasma exists
   a. inside the sun?
   b. in a controlled nuclear fusion device?
   c. in a low-pressure plasma for processing of solids?

1.8 REFERENCES

[1] Langmuir, I., Phys. Rev., **33**: 954 (1929).

[2] Cecchi, J. L., in *Handbook of Plasma Processing Technology*, eds. Stephen M. Rossnagel, Jerome J. Cuomo, and William D. Westwood, p. 14. Park Ridge, NJ: Noyes Publications, 1990.

[3] Thornton, J. A, and A. S. Penfold, in *Thin Film Processes*, eds. J. L. Vossen and W. Kern, p. 75. New York: Academic Press, 1978.

[4] Chen, F. F., *Introduction to Plasma Physics and Controlled Fusion,* 2nd ed. New York: Plenum Press, 1984.

[5] Venugopalan, M., ed. *Reactions Under Plasma Conditions*, Vol. I. New York: Wiley-Interscience, 1971.

[6] Bell, A. T., in *Techniques and Applications of Plasma Chemistry*, eds. John R. Hollahan and Alexis T. Bell, p. 1. New York: John Wiley & Sons, 1974.

[7] Bell, A., in *The Application of Plasmas to Chemical Processing*, eds. Raymond F. Baddour and Robert S. Timmins. Cambridge, MA: MIT Press, 1967.

[8] Chapman, I. B., *Glow Discharge Processes: Sputtering and Plasma Etching.* New York: John Wiley & Sons, 1980.

[9] Chen, F. F., in *Plasma Diagnostic Techniques*, eds. Richard H. Hudlestone and Stanley L. Leonard, p. 113. New York: Academic Press, 1965.

[10] Hershkowitz, N., in *Plasma Diagnostics,* eds. O. Auciello and D. L. Flamm, Vol. I, p. 113. New York: Academic Press, 1989.

[11] Vandertop, G. J., M. Kawasaki, R. M. Nix, I. G. Brown, M. Salmeron, and G. A. Samorjai, Phys. Rev., B**41**, 3200 (1990).

[12] Vossen, J. L., and W. Kern, eds. *Thin Film Processes.* New York: Academic Press, 1978.

[13] Brown, S. C., *Basic Data of Plasma Physics*. Cambridge, MA: MIT Press, 1959.

[14] Cordey, J. G., R. J. Goldston, and R. R. Parker, *Phys. Today*, **45**: 22 (1992).

2

Cold Plasma Generation

The cold plasmas are usually excited and sustained electrically by direct current (DC), radio frequency (RF), or microwave (MW) power applied to a gas. Plasma chemistry in cold plasmas is controlled mainly by electron energies and gas temperatures. Therefore, as far as identical energies and temperatures can be achieved, the type of discharge used to create the plasma is of little importance. The choice of a specific method and equipment to produce discharges is determined by requirements of flexibility, process uniformity, cost, and process rates. The different methods used for the generation of cold plasmas are described in the following sections, while Chapter 4 describes the reactors used in plasma processing.

2.1 DC GLOW DISCHARGES

A DC glow discharge is produced by applying a DC voltage between two conductive electrodes inserted into a gas at low pressure as illustrated in Fig. 2-1. A high-impedance power supply is used to provide the electrical field.

A small amount of free electrons is always present in the gas, as a result of ionization by naturally occurring radioactivity or cosmic rays. Free electrons can also be produced by photoionization or field emission. As the voltage applied to the gas in the discharge tube is gradually increased, the available free electrons are accelerated in the electric field, thereby gaining kinetic energy. Concomitantly, the electrons lose energy in collisions with the atoms or molecules of the gas. These atoms or molecules will be referred to also as collision targets.

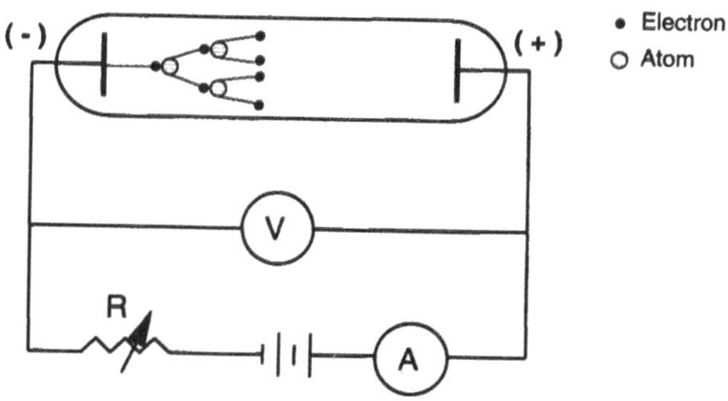

Fig. 2-1 DC glow-discharge setup.

Initially, when the energy of the electron is too low to excite or ionize a target, the collisions will necessarily be elastic. The average fraction of the electron energy lost in an elastic collision with a gas atom or molecule is $-2m_e/M$, where M is the mass of the target (Eq. 1.7 in Sec. 1.3). Thus only a very small fraction of the total kinetic energy of the electron, typically only 10^{-5}, is lost per elastic collision. Meanwhile, the electron continues to gain energy between collisions until it attains sufficient energy to cause ionization of the targets through inelastic collisions. Large amounts of energy are transferred to the target in the inelastic collisions, making those collisions a more efficient mean of energy transfer. The new electrons produced in the ionization process are in turn accelerated by the electric field and produce further ionization by impact with the neutral atoms or molecules of the gas.

An *electron multiplication* process thus takes place. This process can be characterized by a macroscopic coefficient, a_T, which represents the mean number of ion-electron pairs formed by an electron along a path of 1 cm. The coefficient a_T is called the *first Townsend coefficient* and is dependent on the electric field (E), *the pressure* (p), and the nature of the gas as well. Experimentally, it has been shown that for a given gas, a_T/p depends on E/p [1] only.

The changes that take place in the gas as a function of the applied voltage are described by the typical dependence of the discharge current, I, on the applied voltage, V, This dependence is called the *I-V characteristic* of the discharge. A representative typical characteristic of a DC glow discharge is shown in Fig. 2-2. When the applied voltage is low, the current through the tube is produced by the collection of the available free charges and is negligibly small. When the voltage increases and more charged particles are created by the ionization of the gas, the current increases steadily, while the voltage reaches a limit determined by the output impedance of the power supply. This region of the *I-V* characteristic is known as the *Townsend discharge* and indicated in Fig. 2-2.

When the applied voltage reaches a certain threshold value, marked V_b in Fig. 2-2, an avalanche process occurs as a result of three simultaneous processes:

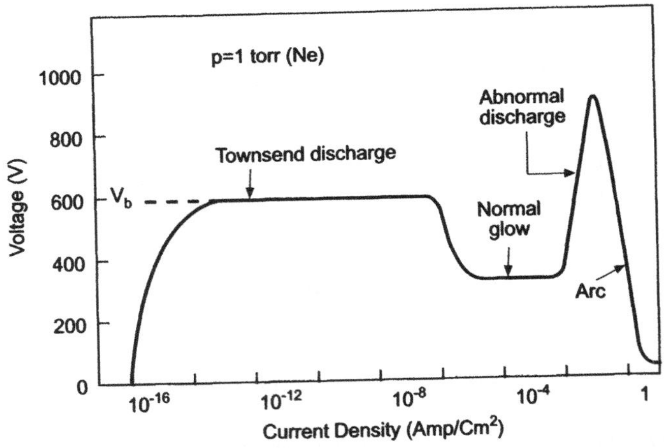

Fig. 2-2 The I-V characteristic of a DC glow discharge.

1. Ions accelerated by the electric field strike the cathode with sufficient energy to cause the *emission of secondary electrons* from it; the secondary electrons in turn form more ions by collision with the neutral atoms of the gas.
2. The newly formed ions are accelerated to the cathode where collisions produce more electrons, which in turn produce more ions.
3. In parallel, electrons created in ionizing collisions and by secondary processes are removed from the plasma by drift and diffusion to the walls, by recombination with positive ions, and, in certain gases, by formation of negative ions. At pressures lower than a few mtorr, the recombination occurs mainly at the walls of the reactor.

When the number of electrons is sufficient to produce just enough ions to regenerate the number of lost electrons, a steady state is reached in which an equilibrium is established between the rate of formation of ions and the rate of their recombination with electrons. At this stage the discharge is self-sustaining. Extensive breakdown occurs in the gas and the *glow discharge* is thus established. The gas begins to glow, the voltage drops, and the current rises abruptly. The mode of the discharge at this point is called the *normal glow* (see Fig. 2-2). The energy decay of the electronically excited states of molecules and atoms account for much of the luminous glow of the gaseous discharge.

The minimal threshold voltage required to produce the glow discharge, V_b, is called the *breakdown voltage*.

The sustaining glow discharge is controlled by the emission of secondary electrons. That emission is characterized by the *electron emission coefficient*, defined as the ratio between the number of emitted secondary electrons and the number of impacting ions. This coefficient is often called the *yield of secondary electrons*. Because the electron emission coefficient is of the order of 0.1 for most materials, much more than one ion must strike the cathode to produce one electron. Then, depending on losses, each secondary electron must be responsible for 10–20 ionizations to sustain a continuous stable plasma.

Although ion impact on the cathode is the main source of secondary electrons in a DC glow discharge, other mechanisms can also contribute to secondary electron emission. Among those, the most dominant are:

1. Ionization of gas atoms or molecules by the photons emitted during ion-electron recombination (*photoionization*)
2. Emission of secondary electrons by the cathode due to impact of photons (*photoemission*)
3. Emission of secondary electrons by the cathode due to impact of excited atoms in a metastable state

The contribution of each of these effects to the emission of secondary electrons is strongly dependent on the nature of the gas and its pressure.

After the voltage reaches the breakdown value V_b and the glow discharge is established, initially at low power, the discharge covers only the area near the rim of the cathode. When the power is raised, the current increases and the discharge spreads to cover the whole surface of the cathode. When the current of the discharge increases past the point of complete cathode coverage, the voltage begins to rise. This corresponds to the *abnormal discharge* region in Fig. 2-2, which is the mode used in glow discharge processing [2]. At this stage, the discharge current is limited by the surface area of the electrodes and by the resistance of the power supply and electrical circuit. A further increase in power will cause the heating of the cathode, which will, at a certain stage, result in thermionic emission. When this happens, the voltage decreases and the glow changes into an arc.

2.1.1 Paschen's Law

The breakdown voltage in a DC discharge is determined by the discharge gas, the gas pressure, and the tube dimension. The dependence of the breakdown voltage on gas pressure and inter-electrode distance can be expressed as [3]:

$$V_b = \frac{C_1\,(pd)}{C_2 + \ln(pd)} \tag{2.1}$$

where d = distance between electrodes

C_1 and C_2 = constants that change with the nature of the gas

According to Eq. (2.1), for large pd values the breakdown voltage is proportional to pd. This dependence is called *Paschen's law*, which reflects in fact the dependence of the DC glow discharge on the secondary electrons.

If the distance between the electrodes is small or the pressure is small, the secondary electrons emitted from the cathode can reach the anode while undergoing only a very small number of collisions not creating a sufficient number of ions required for the regeneration of the secondary electrons. On the other hand, if the pressure is too high, the electrons cannot acquire sufficient energy between collisions to produce enough ions. If the distance between electrodes is too large, only a small fraction of the produced ions will succeed in reaching the cathode and create secondary electrons.

At both extremes of the value of the *pd* product, the probability of ionization — and / or ion collection — is small and the breakdown voltage required to sustain the discharge is high. The breakdown voltage reaches a minimum between the two extremes. This behavior is described by Paschen's curves, which are shown for a few gases in Fig. 2-3 [4].

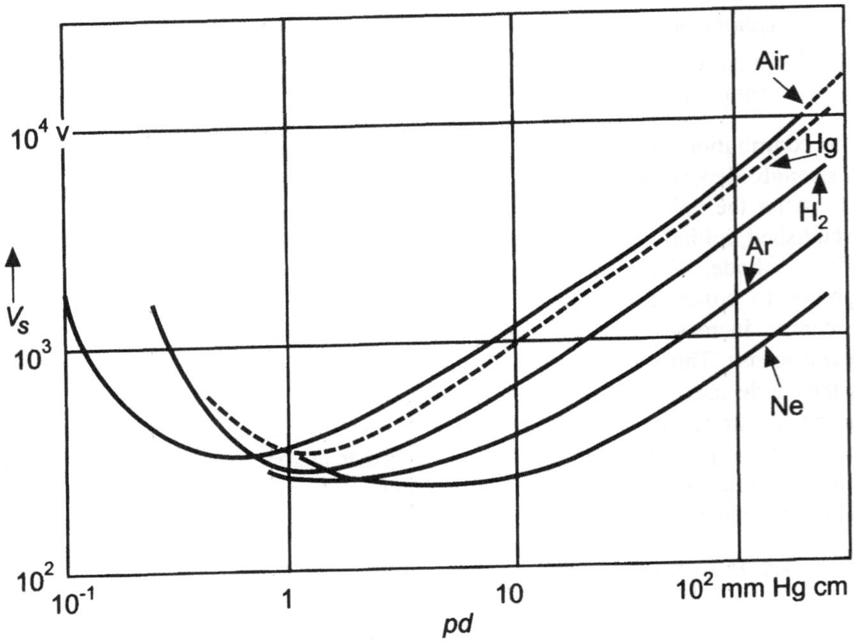

Fig. 2-3 Paschen's curves for various gases (from [4], by permission of Oxford University Press).

For most gases the minimum breakdown voltage is between 100 and 500 V and occurs for *pd* in the range of 10^{-1}–10 torr.cm. At 1 torr, typical values of the electric field are 10–100 $V.cm^{-1}$. Impurities in the gas can reduce the breakdown voltage, either through Penning ionization (see Sec. 3.3.1.2.) or when they have a low ionization potential.

2.1.2 Characteristics of DC Glow Discharge

Figure 2-4 shows a diagram of the regions of the discharge as they appear in the discharge tube together with the distribution of the potential and electric field along the tube. As can be seen in Fig. 2-4 (a), the discharge exhibits several bright and dark regions along the discharge tube. The dark regions are called the *cathode* or *Crooke's dark space*, the *Faraday dark space*, and the *anode dark space*. The luminous regions are called the *cathode glow*, the *negative glow*, and the *positive column*.

The luminosity of the cathode glow results from the decay of the excitation energy of the positive ions during neutralization with electrons.

The cathode dark space is a positive space charge plasma sheath. At a given applied discharge voltage the product between the thickness of the cathode dark space, d_{cd}, and the pressure in the discharge, p, is approximately constant. For example, for a discharge in argon at 2000 V, $p \cdot d_{cd} = 50$ mtorr.cm. The secondary electrons produced at the cathode by positive ion bombardment gain energy through the cathode dark space.

Most of the glow discharge volume is occupied by the positive column.

The voltage drop occurs over three regions indicated in Fig. 2-4(b): the *cathode fall*, the *positive column*, and the *anode fall*. Most of the voltage drop occurs in the cathode fall region, whose dimension is a function of the material used for the cathode and the nature of the discharge gas. Values of the voltage drop in the cathode fall are presented in Table 2-1 for different combinations of cathode material and discharge gas. As can be seen from the table, the voltage values range from 59 V to about 500 V.

Since at constant voltage the product of the thickness of the dark space with pressure is fairly constant, the distance over which the cathode fall extends decreases with increasing pressure. The value of the product $p \cdot d_{cd}$ is presented in Table 2-2 for several combinations of cathode material and discharge gas. For the pressures normally used in DC glow discharges, these values correspond to dark space thicknesses, d_{cd}, of order of millimeters to centimeters.

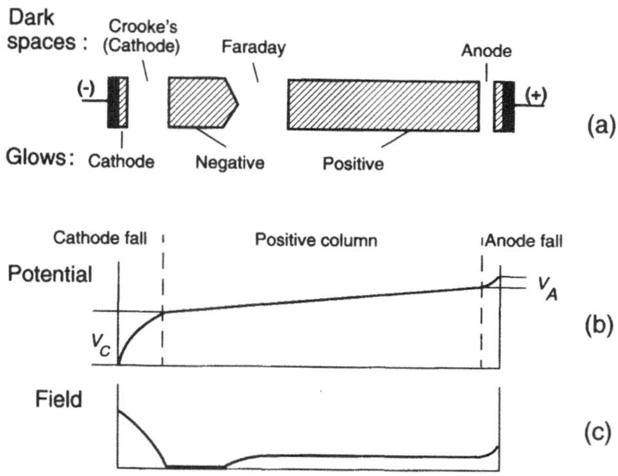

Fig. 2-4 Regions and characteristics of a DC glow discharge: (a) discharge regions; (b) potential distribution in discharge tube; (c) distribution of electric field in discharge tube.

TABLE 2-1 Values of Cathode Fall (in volts) (from [5], reprinted with permission from S. C. Brown, *Basic Data of Plasma Physics*, MIT Press, 1959)

Cathode	Air	A	He	H_2	Hg	Ne	N_2	O_2	CO	CO_2	Cl
Al	229	100	140	170	245	120	180	311	—	—	—
Ag	280	130	162	216	318	150	233	—	—	—	—
Au	285	130	165	247	—	158	233	—	—	—	—
Ba	—	93	86	—	—	—	157	—	—	—	—
Bi	272	136	137	240	—	—	210	—	—	—	—
C	—	—	—	240	475	—	—	—	525	—	—
Ca	—	93	86	—	—	86	157	—	—	—	—
Cd	266	119	167	200	—	160	213	—	—	—	—
Co	380	—	—	—	—	—	—	—	—	—	—
Cu	370	130	177	214	447	220	208	—	484	460	—
Fe	269	165	150	250	298	150	215	290	—	—	—
Hg	—	—	142	—	340	—	226	—	—	—	—
Ir	380	—	—	—	—	—	—	—	—	—	—
K	180	64	59	94	—	68	170	—	484	460	—
Mo	—	—	—	—	353	115	—	—	—	—	—
Mg	224	119	125	153	—	94	188	310	—	—	—
Na	200	—	80	185	—	75	178	—	—	—	—
Ni	226	131	158	211	275	140	197	—	—	—	—
Pb	207	124	177	223	—	172	210	—	—	—	—
Pd	421	—	—	—	—	—	—	—	—	—	—
Pt	277	131	165	276	340	152	216	364	490	475	275
Sb	269	136	—	252	—	—	225	—	—	—	—
Sn	266	124	—	226	—	—	216	—	—	—	—
Sr	—	93	86	—	—	—	157	—	—	—	—
Th	—	—	—	—	—	125	—	—	—	—	—
W	—	—	—	—	305	125	—	—	—	—	—
Zn	277	119	143	184	—	—	216	354	430	410	—

The voltage drop at the anode fall is of the same order as the ionization potential of the discharge gas and usually extends over a few millimeters, as illustrated in Table 2-3. In the anode region secondary electrons are created primarily from electron bombardment. These are accelerated in the anode fall and are a source of particles and energy.

Most of the ionization needed to supply the ions that are accelerated to the cathode takes place in the cathode sheath. However, high electric fields extend into the negative glow, causing ionization also in the negative glow region.

Chemical reactions occur mainly in the cathode fall and positive column zones.

When the distance between the electrodes in a DC discharge decreases, the Faraday dark space and positive column shrink and finally disappear, leaving only the negative glow and cathode dark space. This is generally the situation encountered in most glow-discharge processing reactors. The minimum distance between

TABLE 2-2 Values of the Product Cathode Dark Space × Pressure (in torr.cm) (from [5], reprinted with permission from S. C. Brown, *Basic Data of Plasma Physics*, MIT Press, 1959)

Cathode	Air	A	H_2	He	Hg	N_2	Ne	O_2
Al	0.25	0.29	0.72	1.32	0.33	0.31	0.64	0.24
C	—	—	0.9	—	0.69	—	—	—
Cd	—	—	0.87	—	—	—	—	—
Cu	0.23	—	0.8	—	0.6	—	—	—
Fe	0.52	0.33	0.9	1.30	0.34	0.42	0.72	0.31
Mg	—	—	0.61	1.45	—	0.35	—	0.25
Hg	—	—	0.9	—	—	—	—	—
Ni	—	—	0.9	—	0.4	—	—	—
Pb	—	—	0.84	—	—	—	—	—
Pt	—	—	1.0	—	—	—	—	—
Zn	—	—	0.8	—	—	—	—	—

TABLE 2-3 Anode Fall Voltage, V_A, and Anode Fall Thickness, d_A (after [5], reprinted with permission from S. C. Brown, *Basic Data of Plasma Physics*, MIT Press, 1959)

Gas	Pressure (torr)	V_A (V)	d_A (cm)
Hydrogen	1–5	17–19	0.5–0.7
Nitrogen	0.8	16.5	
Nitrogen	0.5	15.7	
Oxygen	1	14.2	
Oxygen	~0.1		>0.12
Helium	10	26	
Neon	10	17	
Argon	20	10	
Argon	0.5	15.3	

the electrodes has to be about twice the thickness of the cathode dark space; at smaller electrode separation, the discharge is extinguished [6].

2.2 RADIO FREQUENCY DISCHARGES

To sustain a DC discharge, electrically conductive electrodes have to be inserted inside a reactor and be in direct contact with the plasma. If the discharge is used for deposition of dielectric films, the electrodes exposed to the plasma gradually become covered with an insulator. Therefore, although a DC discharge may be initiated, it will quickly extinguish as the electrons accumulate on the insulator and recombine with the available ions. In some cases it is preferred to have the electrodes outside the reactor, to avoid or minimize contamination of the process by material removed from the electrodes. Such problems can be solved by alternating the polarity of the discharge.

When an alternating electric field of low frequency (< 100 Hz) is applied between the two electrodes of the discharge tube, each electrode acts alternately as cathode or anode. Once the breakdown potential is surpassed on each half cycle, a temporary DC glow discharge is obtained. When the voltage drops during the cycle below the breakdown value, the discharge is extinguished, and for sufficiently low frequencies, the space charge decays before the discharge is reinitiated with inverse polarity [7].

When the frequency of the electric field increases above a *critical ion frequency*, f_{ci}, defined by Eq. (2.2), the time taken by the positive ions to move between electrodes becomes larger than half the period of the electric field. Ions created near a momentary anode cannot reach the cathode before the field is reversed. In this situation, the distance traveled by the ions in the electric field becomes smaller than the thickness of the plasma sheath. At such frequencies the positive space charge is partly retained between the two half cycles of the alternating electric field and facilitates the reinitiation of the discharge. The critical ion frequency, also called the *ion transition frequency* [8], is defined by [9]:

$$f_{ci} = \frac{< v >_{di}}{2L} \tag{2.2}$$

where $< v >_{di}$ = average drift velocity of the ion
L = the distance between the two electrodes

Ion transition frequencies range from 500 kHz to several MHz.

A similar critical frequency can be defined for electrons:

$$f_{ce} = \frac{< v >_{de}}{2L} \tag{2.3}$$

where $< v >_{de}$ is the average drift velocity of the electrons.

Due to the much larger mobility of the electrons as compared to that of the ions, f_{ce} is much higher than f_{ci}. For frequencies higher than f_{ce}, both positive and negative space charges are retained between cycles, and, as a result, the voltages required to initiate and maintain the alternating current discharge decrease strongly in comparison to a DC glow discharge [5, 9].

The frequencies used in the high-frequency discharges are in the range of radio transmission giving the high-frequency discharges the name of *radio frequency*, or shortly, *RF discharges*.

The elastic collision frequency, ν, in gases at glow discharge conditions is normally between 10^9 and 10^{11} collisions/sec [10]. Thus the collision frequency is much higher than the applied radio frequency even for 13.56 MHz discharges and electrons will experience many collisions during each applied field cycle. They will be lost by diffusion to the reactor walls and will be regenerated by impact ionization in the body of the plasma. Therefore, the loss of electrical carriers from the RF discharge is controlled by ambipolar diffusion and homogeneous recombination (recombination in the gas phase) and not by the electric field. New charged particles are produced mainly through electron impact ionization of neutral gas atoms and molecules.

The *power absorption* by the RF discharge can be either collisional or collisionless. The collisional absorption of high-frequency power in the plasma is due to collisions of electrons with ions, at a frequency ν_{ei}, and with neutral particles, at a frequency ν_{en}. In plasmas at pressures above 10^3 Pa (7.5 torr), the degree of ionization is usually very low ($< 10^{-4}$), the density of neutrals is much higher than that of ions, and the electron-neutral collisions are predominant. At pressures below 1 Pa (7.5 mtorr), the degree of ionization can reach values higher than 10^{-2} (see Table 1-1 in Sec. 1.3.1), and the electron-ion collisions are predominant. At intermediate pressures, both ν_{ei} and ν_{en} determine the power absorption [11]. At even lower pressures, when the collision frequency decreases and $\nu/\omega \ll 1$, collisionless absorption becomes dominant in the plasma.

In a collisionless situation, an electron would oscillate in the RF field and would reach maximal velocity $\dot{x}$, amplitude x, and energy W, given by [6]

$$\dot{x} = \frac{eE_o}{m_e\omega} \tag{2.4a}$$

$$x = \frac{eE_o}{m_e\omega^2} \tag{2.4b}$$

$$W = \frac{m_e\dot{x}^2}{2} \tag{2.4c}$$

where E_o is the amplitude of the electric field.

At a typical RF frequency of 13.56 MHz and a field strength of 10 V/cm, this corresponds to

amplitude = 2.42 cm; velocity = 2.1×10^8 cm.sec^{-1};

energy = 11.3 eV

This indicates that, for an electron to reach the ionization energy of argon (15.7 eV) in a collisionless plasma, a field somewhat higher than 10 V/cm is required. However, the collisions with the atoms of the gas cause a random motion of the electrons and the electrons acquire additional energy from the external field during each collision with the atoms. If an electron makes an elastic collision with an atom, reversing its motion at the time the electric field changes direction, it will continue to gain speed and energy [12]. Electrons in a RF discharge could thus accumulate enough energy to cause ionization even at low electric fields. As a result of this behavior, the RF discharge is more efficient than the DC discharge in promoting ionization and sustaining the discharge.

The *mean power absorbed* by an electron, $\bar{P}$, is given by [1]

$$\bar{P} = \frac{e^2E_o^2}{2m_e} \cdot \frac{\nu_{ea}}{\nu_{ea}^2 + \omega^2} \tag{2.5}$$

where ν_{ea} is the elastic collision frequency of an electron with the atoms or molecules of the gas. The dependence of $\bar{P}$ on E_0^2 indicates that the absorbed power is independent of the sign of the electric field and the electron gains energy both when it moves with the field or against it. The quantity

$$E_{eff} = \frac{E_o}{\sqrt{2}} \left(\frac{\nu_{ea}^2}{\nu_{ea}^2 + \omega^2} \right)^{1/2} \tag{2.6}$$

is called the *effective electric field strength.*

The *average RF power* transferred from the outside electric field to the *unit volume of gas*, $\bar{P}_v$, is

$$\bar{P}_v = \frac{n_e e^2 E_0^2}{2m_e} \left(\frac{m}{\nu^2 + \omega^2} \right) \tag{2.7}$$

For RF frequencies of $\omega \approx 10^7$ Hz, and with $\nu > 10^9$ $\sec^{-1}$, the collision frequency is much higher than the frequency of the field $\nu \gg \omega$. In this case the transferred power, $\bar{P}_v$, is practically not affected by the driving frequency, ω.

At frequencies higher than 50 kHz, the oscillating electrons acquire sufficient energy to cause ionizing collisions, thus reducing the dependence of the discharge on secondary electrons and lowering the breakdown voltage. The discharge can thus be sustained independent of the yield of secondary electrons from walls and electrodes. Because of the low mobility of the ions, their distribution is essentially stationary at high frequencies. In the same time, the electrons are swept between the electrodes by the electric field and their distribution is a function of position and time.

The secondary electrons emitted from the electrodes or walls of the reactor of an RF discharge are accelerated across the plasma sheath and add to the ionization process but are less important in sustaining the discharge. At a given pressure, the electrical impedance of the discharge decreases with increasing frequency, making it possible to drive more current through the discharge at the same voltage [6]. The yield of atoms and free radicals in a molecular discharge is also increased by the use of RF excitation as compared to a DC discharge of the same field strength and pressure.

Although the frequency of power supplies used for the excitation of RF plasmas can be as high as 100 MHz, lower frequencies are generally used. This is due to the fact that sophisticated arrangements have to be made to sustain a uniform plasma in large volumes when the wavelength becomes comparable to, or shorter than, the dimensions of the reactor.

The RF discharges can be operated at pressures as low as 1 mtorr because the efficiency of ionizing collisions is enhanced by the electron oscillations. This is useful in sputtering, where it is undesirable to have sputtered material reflected back to the etched surface as a result of collisions with gas molecules, or in etching or deposition, when directionality of the ions is required.

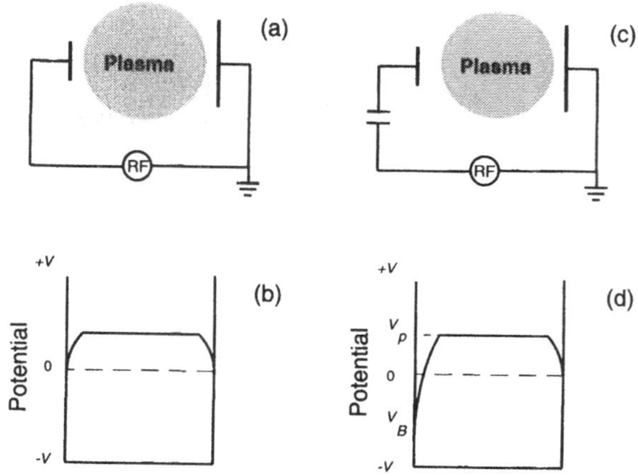

Fig. 2-5 RF connection to parallel plate reactor with unequal electrodes and developed self-bias: (a) direct RF connection to electrodes; (b) potential distribution between electrodes for arrangement (a); (c) RF connection through blocking capacitor; (d) potential distribution between electrodes for arrangement (b).

2.2.1 Self-bias in RF Plasmas

Let's consider a RF plasma generated between two parallel electrodes, as shown in Fig. 2-5, and assume that one electrode has an area much larger than the other. The electrodes will be at negative potentials, V_1 and V_2, relative to the plasma and the sheaths of thickness d_{s1} and d_{s2} will develop near the two electrodes of areas A_1 and A_2. If the RF field is connected directly to the electrodes, as illustrated in Fig. 2-5 (a), the two electrodes will be at the same potential relative to the plasma (Fig. 2-5 (b)), because the plasma is equipotential:

$$V_1 = V_2 \tag{2.8}$$

$$d_{s1} = d_{s2} \tag{2.9}$$

The situation changes, however, if a blocking capacitor is inserted, as often happens, between the RF supply and the electrodes, as shown in Fig. 2-5 (c). In this case, the potential distribution between the electrodes is as illustrated in Fig. 2-5 (d) and this nonsymmetric potential develops as explained next.

If a square-wave potential of amplitude V, as illustrated in Fig. 2-6 (a), is applied to the electrodes through the capacitor, the voltage across the plasma will follow the curve shown in Fig. 2-6 (b) [3]. Initially the voltage across the plasma will be equal to the applied voltage V. The capacitor will charge up rapidly by electron current, and the potential will drop as shown in Fig. 2-6 (a). When the applied voltage changes sign, the voltage across the plasma drops by -2 V, after which the voltage decays at a slower rate because the capacitor charges this time by the current of less mobile ions. This process continues until the time-averaged ion and electron currents become equal.

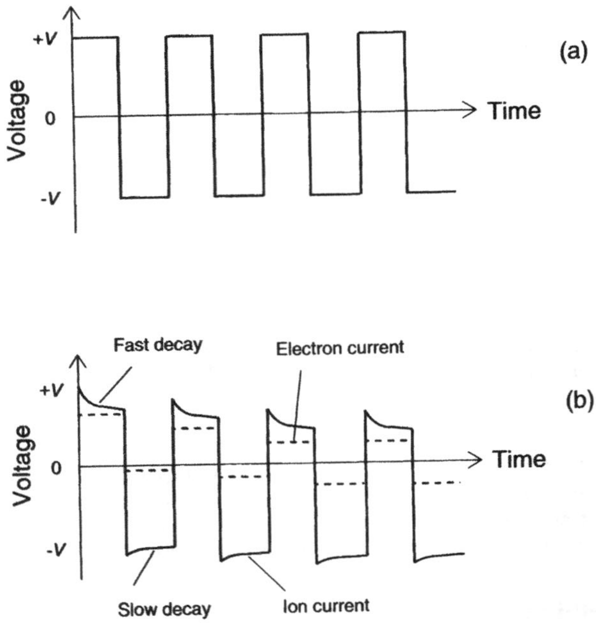

Fig. 2-6 Development of a self-bias in a parallel plate discharge: (a) applied voltage; (b) voltage across the discharge versus time.

This steady state results in a *time-averaged negative bias* on the small electrode. The time-averaged potential distribution between the electrodes is therefore as illustrated in Fig. 2-5 (d). The same will happen if a sinusoidal voltage is applied to the electrodes, as illustrated in Fig. 2-7 (a). This figure also illustrates the time variation of the plasma potential (curves 2) for the symmetric and nonsymmetric cases. If the electrodes have the same area, the system remains symmetric, as shown in Fig. 2-7 (b).

In all cases, the plasma is at a higher potential than each electrode.

When a blocking capacitor is used with two electrodes of different areas, the negative self-bias of the electrodes relative to the plasma is dependent on the relative areas of the two electrodes. According to Koenig and Maissel [13], the following relation exists,

$$\frac{V_1}{V_2} = \left(\frac{A_2}{A_1}\right)^4 \tag{2.10}$$

$$\frac{d_{s1}}{d_{s2}} = \left(\frac{V_1}{V_2}\right)^{3/4} \tag{2.11}$$

where V_1, V_2 = negative biases on the two electrodes
A_1, A_2 = areas of the electrodes

Often, one of the electrodes is grounded together with the walls of the reactor and

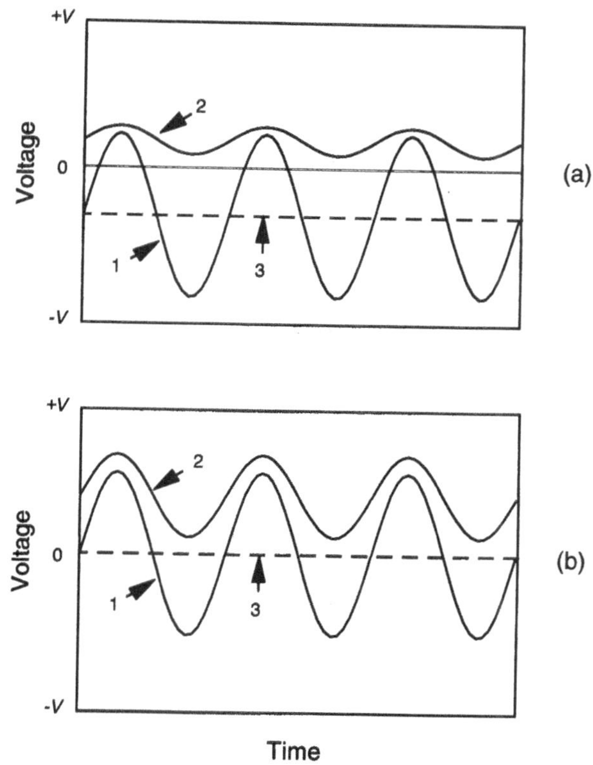

Fig. 2-7 Development of a self-bias with a sinusoidal wave: (a) unequal electrodes; (b) electrodes of equal area.
1, voltage on powered electrode; 2, plasma potential; 3, time-averaged voltage on powered electrode.

its effective area becomes very large, causing the other electrode to become much more negative.

Equation (2.10) was obtained assuming that the ions pass through the sheath without collisions and that the current density of the ions is equal at the two electrodes. The first assumption is correct only at pressures of a few mtorr, and the second assumption is questionable. Equation (2.10) takes into consideration the total areas of the electrodes, which includes the area of the reactor for the electrode that is electrically connected to it. However, the plasma is often mostly confined between the electrodes while the walls are practically not exposed to it.

The experimental evidence indicates that the voltage ratio V_1 / V_2 depends also on the gas used in the discharge, the peak-to-peak applied voltage, as well as the area ratio $(A_2/A_1)^n$. For area ratios between 0.6 and 0.1 the fourth power relationship is approximately obeyed. However, it is not obeyed for smaller area ratios. For an Ar discharge with an electrode area ratio of 0.3, the value of n can range from 1.2 to 2.5 [14].

A RF power supply is connected to a plasma reactor through an *impedance matching unit.* The unit is needed to match the impedance of the plasma reactor to the output impedance of the power supply which is usually at 50 Ω. Because the matching units normally do include a blocking capacitor in the planar diode reactors used in PECVD processing, the electrodes are made of equivalent areas when similar voltages relative to the plasma are required on both electrodes. However, the described effect can be exploited to obtain a higher negative bias on one electrode.

The value of the self-bias is dependent on the RF power applied to the electrode and the pressure in the reactor. It increases with increasing power and decreasing pressure according to [15]

$$V_B \propto \left(\frac{P_{RF}}{p} \right)^{1/2} \tag{2.12}$$

where V_B = self-bias on the powered electrode
P_{RF} = RF power

As the frequency of the electric field increases (e.g., from 50 kHz to 13.56 MHz), there is increasingly less time between cycles available for the diffusion of the charged particles to the reactor walls. Therefore, according to Bohm sheath criterion (see Sec. 1.3.4), less negative bias has to develop across the sheath to keep the electrons in the plasma. This, in turn, means that the ion bombardment of the electrode surface decreases with increasing frequency of the electric field [14].

If each electrode shown in Fig. 2-5 is RF powered relative to the grounded chamber, it is possible to control the DC bias on each electrode independently. Thus the plasma can be generated by powering one electrode and the degree of ion bombardment on the other electrode can be regulated independently by a second power supply.

Magnetic fields parallel to the electrode are sometimes superposed on the RF plasma. The magnetic field confines the electrons next to the electrodes and increases the ionization efficiency. This increase in ionization lowers the sheath potential and the energy of the bombarding ions.

2.2.2 RF Versus DC Plasmas

The advantages of RF discharges over the DC discharges, as summarized next, explain the wider use of RF plasmas as compared to DC plasmas:

1. RF plasmas can be excited and sustained using either conductive or nonconductive electrodes, while DC discharges require the electrodes to be conductive throughout the process.
2. RF plasmas can be sustained with internal as well as external electrodes, while DC discharges require the electrodes to be inserted inside the reactor and be in direct contact with the plasma. Use of external electrodes is sometimes required when the gases of the discharge are corrosive or when

one wants to reduce contamination of the plasma with the material of the electrodes.

3. RF plasmas are characterized by higher ionization efficiencies than are the DC plasmas.
4. RF plasmas can be sustained at lower gas pressures than are DC plasmas.
5. In RF plasmas the energy of the ions bombarding the sample is controlled by the negative bias, which can be adjusted over a wide range of values. Samples placed on the cathode of the DC discharge are exposed to bombardment of high-energy ions that are accelerated at voltages that have to be above the minimal breakdown voltage. This can cause damages to sensitive substrates.

2.3 MICROWAVE PLASMAS

Microwave plasmas are sustained by power supplies operating at a frequency of 2.45 GHz. This frequency, which is commonly used for industrial or home heating applications, makes suitable power supplies readily available. The excitation of the plasma by microwaves is similar to the excitation with RF, while differences result from the ranges of frequencies.

In a typical microwave plasma the strength of the electric field is about $E_o \approx 30$ V/cm; therefore, according to Eqs. (2.4a)–(2.4c), in a collisionless situation, the maximum amplitude of the electron at microwave frequencies is $x < 10^{-3}$ cm, and the corresponding maximum energy acquired by an electron during one cycle is about 0.03 eV. This energy is far too small to sustain a plasma. Therefore microwave discharges are more difficult to sustain at low pressures (< 1 torr) than DC or RF discharges.

In a collisional discharge, at constant electric field and power density, Eq. (2.7) has a maximum when $\nu = \omega$. The absorption of microwave power is thus a function of the collision frequency of the electrons with the heavy species and is therefore dependent on the pressure in the discharge. For a microwave frequency of 2.45 GHz, efficient microwave absorption in helium occurs at 5–10 torr [16]. For other gases, the optimum pressure for microwave discharges is in the range 0.5–10 torr.

While the RF glow discharge can be made to extend virtually throughout the entire reactor, whose dimensions are much smaller than the wavelength of the RF field (≈ 22 m at 13.56 MHz), the microwave plasma has its greatest glow intensity at the coupling microwave cavity and diminishes rapidly outside it, because of the much smaller wavelength of the microwave ($\lambda = 12.24$ cm for a frequency of 2.45 GHz). In a microwave plasma, the magnitude of the electric field can vary within the reactor, which now has dimensions of the same order of magnitude as the wavelength. One can thus find active species from the discharge still persisting into a region free of the glow of the plasma, that is, in the afterglow.

The transfer of the microwave power to the plasma is achieved with the help of *microwave applicators* such as waveguides, resonance cavities, or coaxial applicators. With the exemption of coaxial applicators, the applicator has to be separated from the plasma reactor by a dielectric wall, characterized by low absorption of microwaves. A window made of quartz or alumina separates the microwave from the plasma. Any absorbing material inserted in the path of the microwave will cause both absorption and reflection of the microwaves, reducing the efficiency of the power transfer and impeding the tuning of the microwave system.

The microwave applicators will be further discussed in detail in Sec. 4.4.

2.4 ELECTRON CYCLOTRON RESONANCE PLASMAS

Electron cyclotron resonance (ECR) plasmas have the capability to operate at lower pressures than RF plasmas and to create higher plasma densities, with corresponding higher degrees of ionization ($a > 10\%$ in some cases [17]).

According to Eq. (2.5), which describes the relation between the absorbed power and the frequency of the electric field, the power absorbed by plasma approaches zero for conditions of rare collisions, when $\nu \ll \omega$, or frequent collisions, when $\nu \gg \omega$. Very high values of electric field, E_o, are therefore required to sustain microwave discharges at very low pressures of less than 1 Pa (7.5 mtorr), for which $\nu \approx 10^8\ \text{sec}^{-1}$.

If a magnetic field B is applied to a plasma system, the charged particles are subjected to a gyromotion around the magnetic field lines with a radius r_L and an angular frequency ω_c. The radius r_L is called the *Larmor radius* and is given by [8]

$$r_L = \frac{mv_\perp}{eB} = \frac{1}{eB}\sqrt{2\frac{W_\perp}{m}} \tag{2.13}$$

where m = mass of charged particle
$v_\perp$ = velocity component of particle, normal to the magnetic field line
$W_\perp$ = energy component corresponding to the normal component of the velocity

The frequency ω_c is called the *cyclotron angular frequency* and is given by

$$\omega_c = \frac{eB}{m} \tag{2.14}$$

As can be seen from Eq. (2.14), the cyclotron or *Larmor frequency* is independent of the velocity of the individual charged particle. The cyclotron frequency of the electron is therefore a characteristic parameter of the system.

For a magnetic field of 875 gauss, the cyclotron frequency of the electrons defined by Eq. (2.14) with $m = m_e$, becomes 2.45 GHz, and if microwaves of frequency $\omega = 2.45$ GHz are used to excite and sustain the plasma, the gyromotion of the electrons is in resonance with the microwaves. This is called the *electron cyclotron resonance* condition, and a plasma excited in the presence of a magnetic field, which satisfies the resonance condition, is called an *ECR plasma.*

At the ECR condition, the component of the electron velocity normal to the magnetic field increases continuously and the electrons follow a spiralling gyrating path along the magnetic field line. The radius of the electron path is limited by collisions with other particles or with the walls or by the electron moving out of the ECR zone.

The plasma is extracted from the ECR chamber toward the sample by a divergent magnetic field, the intensity of which decreases from the ECR chamber toward the sample. The charged particles are accelerated toward lower magnetic field values, and electrons will stream out of the plasma along the magnetic field lines much faster than ions. This causes the formation of an electric field that accelerates the ions in the plasma toward the substrate holder. Because of the ambipolar diffusion that keeps the plasma neutral, the ions are extracted from the plasma with the same diffusion coefficient as the electrons. The electrical potential is a function of the gradient of the magnetic field and increases with increasing gradient. The energy of the bombarding ions is about 20 eV [18].

An electric field, propagating parallel to the magnetic field lines and rotating in the plan of the gyromotion of the electron, is called a circularly polarized field. If the field is rotating in the same direction as the gyromotion of the electrons, it is a right-hand circularly polarized field. In a right-hand polarized field, the electrons are in fact exposed to a reduced frequency $(\omega - \omega_c)$ and in this case Eq. (2.7) changes to

$$\overline{P}_v = \frac{n_e e^2 E_o^2}{2m_e}\left[\frac{\nu}{\nu^2 + (\omega - \omega_c)^2}\right] \tag{2.15}$$

In the resonance case, when $\Delta\omega = \omega - \omega_c = 0$, the electrons are continuously accelerated by the magnetic field, and according to Eq. (2.15), the average power absorbed by the electrons is maximal. A drastic gain of power transfer to the plasma is therefore obtained at near-resonance conditions, especially for low-collision frequencies. This is illustrated in Fig. 2-8 which shows the absorbed power as a function of the normalized collision frequency, ν/ω.

For a microwave frequency of 2.45 GHz and a near-resonance condition corresponding to $\Delta\omega/\omega = 0.1$, the maximum power absorption is obtained at 10 Pa (0.075 torr), as indicated in Fig. 2-8. At lower pressures, the absorbed power is much higher than at nonresonant conditions [19].

It has been shown [20, 21] that the energy absorption in an ECR plasma is greater than 70 to 75% even in the 10^{-4} torr pressure range. Asmussen [22] has shown that when gas pressure is increased, there is a transition from ECR heating to collisional heating of the electron gas. This means that the ECR coupling technique is especially useful for low-pressure discharges, approximately below 3 torr.

Propagation of an electric field into a plasma, in the absence of an externally applied magnetic field, requires the fulfillment of the condition

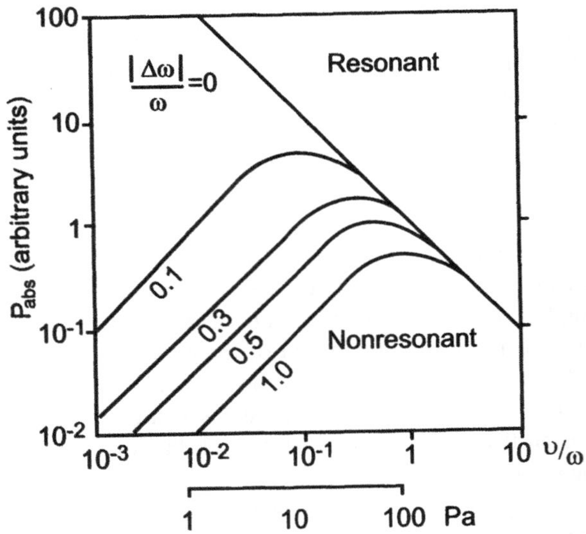

Fig. 2-8 Absorbed microwave power versus normalized collision frequency; the curves correspond to different values of $|\Delta\omega|/\omega$ (from [19], reprinted with permission from *Journal of Vacuum Science and Technology*, vol. A8, p. 908, 1990).

$$\omega \geq \omega_p \tag{2.16}$$

where ω_p is the frequency of the plasma given by Eq. (1.33) from Sec. 1.3.5:

$$\omega_p = \left(\frac{n_e e^2}{m_e \varepsilon_o}\right)^{1/2} \tag{2.17}$$

Eq. (2.16) and Eq. (2.17) can be solved for n_e:

$$n_e \leq \frac{\omega^2 \varepsilon_o m_e}{e^2} = n_c \tag{2.18}$$

Equation (2.18) shows that, for a given frequency ω of the electric field, there is an upper limit n_c for the maximum attainable electron density.

If the electron density, n_e, is too big ($n_e > n_c$) or the field frequency, ω, too small, an electromagnetic wave cannot pass through the plasma. An increase in the plasma density above the critical value will prevent further absorption of the microwave power in a nonmagnetic plasma. The critical density, also called the plasma *cut-off density*, is the highest plasma density achievable in the bulk of nonmagnetic plasmas. For the frequency of 2.45 GHz, the corresponding cut-off density is

$$n_c = 7.1 \times 10^{10} \text{ cm}^{-3} \tag{2.19}$$

However, when a magnetic field, B, is superposed to the plasma, right-hand polarized waves propagate along magnetic field lines, as long as B remains above the ECR value. Plasma densities as high as $n_e = 10^{12}$ cm^{-3} have been obtained in ECR plasmas, when the wave was fed along the magnetic field line [17].

2.4.1 Advantages of ECR Plasmas

The magnetic confinement in the ECR plasmas enhances the degree of ionization, which enables operation at low pressures. ECR plasma sources are usually operated in the pressure range of 10^{-5} to 10^{-3} torr. The combination of low operating pressure with strong microwave coupling generates higher electron temperatures ($T_e \approx 5$ eV) in ECR plasmas as compared to nonmagnetic plasmas. The energy of the ions flowing out of the ECR source is in the range of 10–25 eV, typically three to five times the electron temperature. If necessary, this energy can be increased either by extracting grids at the plasma source or by biasing the processed substrate.

ECR plasmas are becoming a popular etching and deposition tool in microelectronics processing due to their ability to sustain highly dissociated and highly ionized plasmas at very low pressures. Anisotropic etching can be achieved in these plasmas without the damage induced to sensitive substrates by energetic ions in RF plasmas [23].

Some of the advantages of ECR plasmas over RF and DC plasmas are:

1. Lower, broad range of operating pressures
2. High ionization efficiency
3. Wide range of achievable ion energies:
   a. Intrinsic ion energies below values that can induce radiation damage
   b. Higher energies can be achieved easily with assistance of extraction grids, or substrate bias
4. Electrodeless coupling (attractive for use with corrosive gases) of the electric power to the plasma
5. Directionality of the ion and even neutral beam. The directionality of the neutral beam is a result of the large mean free path at low operating pressure and can be exploited for:
   a. Anisotropic etching
   b. Good filling of high-aspect ratio trenches in VLSI processing

2.5 QUESTIONS

1. What are the differences between the plasma regimes corresponding to the Townsend discharge and normal discharge regions of the I-V characteristic of a DC discharge?
2. a. What is the breakdown voltage and what determines its value?
   b. What is the value of the breakdown voltage for a discharge in argon at a pressure of 10 torr if the electrodes are 10 cm apart?
   c. At what pressure will the breakdown voltage be minimal in an argon plasma sustained in the reactor of question 2.(b)?
3. What causes the luminosity of the plasma?

4. What are the advantages of RF plasma excitation as compared to DC excitation? Why can an RF plasma be sustained at lower voltages than required by a DC discharge?
5. a. Assuming an elastic collision frequency of 10^{10} sec^{-1} and a RF field of an amplitude of 10 $V.cm^{-1}$ at 13.56 MHz, calculate the mean power absorbed by an electron in the plasma.

   b. If the plasma density is 10^{10} cm^{-3}, what is the average power transmitted to 1 cm^3 of this plasma?
6. a. Calculate the bias on the electrode of a RF discharge in a cylindrical reactor described in Fig. 2-9 assuming that

   plasma potential = 30 V
   electrode diameter, D_e = 5 in.
   reactor diameter, D_R = 8 in.
   reactor height, h_R = 6 in.

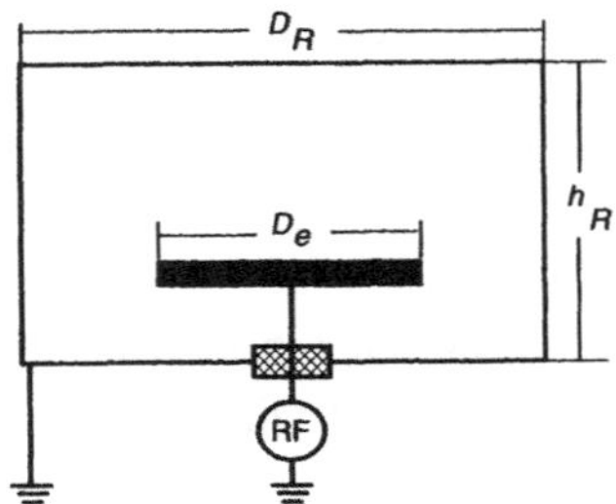

Fig. 2-9 Cylindrical RF reactor for question 6.

   b. What change in RF power is required to double the self-bias on the electrode?
7. a. Calculate the Larmor radius of an electron that has an energy component of 5 eV normal to a magnetic field of 100 gauss.

   b. Calculate the Larmor radius of an argon ion for the same conditions.
8. Calculate the Larmor radius of an argon ion having an energy component of 3 eV normal to the magnetic field at resonance condition in an ECR plasma sustained at 2.45 GHz. How does the Larmor radius determine the dimensions of the reactor and plasma uniformity?
9. What is the maximum electron energy attainable in a microwave plasma sustained at 2.45 GHz? Compare this value to that attainable in a RF plasma at 13.56 MHz.

2.6 REFERENCES

[1] Venugopalan, M., ed., *Reactions Under Plasma Conditions*, Vol. I. New York: Wiley-Interscience, 1971.

[2] Vossen, J. L., and W. Kern, eds., *Thin Film Processes*. New York: Academic Press, 1978.

[3] Cecchi, J. L., In *Handbook of Plasma Processing Technology*, eds. Stephen M. Rossnagel, Jerome J. Cuomo, and William D. Westwood, p. 14. Park Ridge, NJ: Noyes Publications, 1990.

[4] von Engel, A., *Ionized Gases*, p. 195, Oxford, England: Oxford University Press, 1965.

[5] Brown, S. C., *Basic Data of Plasma Physics*. Cambridge, MA: MIT Press, 1959.

[6] Chapman, I. B., *Glow Discharge Processes: Sputtering and Plasma Etching*. New York: J. Wiley & Sons, 1980.

[7] Bell, A. T., In *Topics in Current Chemistry, Plasma Chemistry III*, eds. S. Veprek and M. Venugopalan, p. 43. Berlin: Springer-Verlag, 1980.

[8] Chen, F. F., *Introduction to Plasma Physics and Controlled Fusion*, 2nd ed. New York: Plenum Press, 1984.

[9] Francis, G., *Ionization Phenomena in Gases*. London: Butterworths, 1960.

[10] Baddour, R. F., and R. S. Timmins, eds., *The Application of Plasmas to Chemical Processing*. Cambridge, MA: MIT Press, 1967.

[11] Musil, J., *Vacuum*, 36: 161 (1986).

[12] MacDonald, A. D., and S. J. Tetenbaum, *Gaseous Electronics*, Vol. 1. New York: Academic Press, 1978.

[13] Koenig, H. R., and L. I. Meissel, *IBM J. Res. Dev.*, 14: 276 (1970).

[14] Sherman, A., *Thin Solid Films*, 113: 135 (1984).

[15] Catherine, T., In *Diamond and Diamond-like Films and Coatings, NATO-ASI Series B: Physics*, eds. R. E. Clausing, L. L. Horton, J. C. Angus, and P. Koidl, Vol. 266, p. 193. New York: Plenum Press, 1991.

[16] Asmussen, J., In *Handbook of Plasma Processing Technology*, eds. Stephen M. Rossnagel, Jerome J. Cuomo, and William D. Westwood, p. 285. Park Ridge, NJ: Noyes Publications, 1990.

[17] Holber, W., In *Handbook of Ion Beam Processing Technology. Principles, Deposition, Film Modification and Synthesis*, eds. Jerome J. Cuomo, Stephen M. Rossnagel, and Harold R. Kaufman, p. 21. Park Ridge, NJ: Noyes Publications, 1989.

[18] Matsuo, S., and M. Kiuchi, *J. Appl. Phys.*, 22: L210 (1983).

[19] Geisler, M., J. Kieser, E. Rauche, and R. Wilhelm, *J. Vac. Sci. Technol.*, A8: 908 (1990).

[20] Sakamoto, Y., *Jpn. J. Appl. Phys.*, 16: 1993 (1983).

[21] Mejia, S. R., R. D. McLeod, and K. C. Dao, *Rev. Sci. Instrum.*, 57: 443 (1986).

[22] Asmussen, J., *J. Vac. Sci. Technol.*, A7: 883 (1989).

[23] Hopwood, J., D. K. Reinhard, and J. Asmussen, *J. Vac. Sci. Technol.*, B6: 1896 (1988).

3

Plasma Chemistry

3.1 INTRODUCTION

The initial interest and applications in cold plasmas chemistry were concentrated on synthesizing products in the plasma phase as a final result. Cold plasmas have been studied and used for synthesis of a variety of products in the gas phase, such as ammonia from hydrogen and nitrogen, oxidation of nitrogen, or formation of compounds of inert gases, such as XeF_4. A detailed list of examples can be found in Vursel and Polak [1]. The use of cold plasmas for surface activation of polymers was suggested already in 1968 by Beauchamp and Buttrill [2]. In the mid-1970's, when amorphous hydrogenated silicon was first deposited by PECVD, the direction of research and development started to shift toward the utilization of cold plasma chemistry for deposition of thin films and dry etching, with the microelectronics industry becoming the main user of both applications.

The main advantage of plasma assisted processing of solids evolves from the fact that the chemical reactions can take place at sample temperatures that are considerable lower than possible with thermal reactions taking place at thermodynamic equilibrium. It is, therefore, possible to deposit or grow films on substrates that do not have the thermal stability to accept such coating processes at higher temperatures. In addition, the lower treatment temperatures possible in the plasma are important when one has to limit diffusion or grain growth, especially in microelectronic processing where very small feature sizes and shallow junctions are involved.

In a plasma sustained in a gas at 0.1 to several torr, the electron temperatures reach tens of thousands of °K, while the temperatures of free atoms, molecules, or molecular fragments are only hundreds of °K. The chemical processes occurring in cold plasmas are driven by the energy of the discharge and are not based nor controlled by the temperature of the gas mixture or of the treated surfaces. The chemistry of the cold plasma is therefore a nonthermal, nonequilibrium process. Due to its nonequilibrium nature, the cold plasma has the unique property of efficiently generating chemically reactive species at low temperatures. The nonequilibrium steady-state conditions created by the high-energy electrons can transform a normally inert molecular gas such as carbon tetrafluoride (CF_4) into a highly reactive medium. All species of the cold plasma are essentially potential reactants.

The electrons, produced in the plasma by the partial ionization of the gas, act as the main agent for the transfer of energy from the external electric field to the gas. The energy transfer is done either by elastic or by inelastic collisions between the electrons and the molecules of the gas. Elastic electron-molecule collisions cause an increase in the kinetic energy of the molecules, while the energy transferred in inelastic collisions leads to dissociation and ionization of molecules and formation of plasma species such as free radicals, excited metastables, and ions.

Parallel to the generation of plasma species, a loss process also takes place in the plasma. Electrons are lost mainly by diffusion to the surfaces exposed to the plasma. Ions are similarly lost by diffusion and by recombination with electrons. Neutral atoms and radicals are lost primarily by recombination, through homogeneous reactions in the gas phase, or through surface reactions on the walls of the reactor.

Due to the large number and complexity of elementary reactions encountered in a plasma, it is not intended to give in this book a full account of all the possible reactions, their mechanisms, and kinetics. The differences between the characteristic times of various physical and chemical processes that take place in the cold plasmas are small and do not permit to distinguish between the diverse processes when considering chemical kinetics. In general, more than one reaction can contribute to the production and loss of each species adding to the overall complexity. In certain plasma systems, a single production or loss mechanism may be dominant. Detailed discussion of the reactions in the plasma are available in various monographs [3-5]. Only typical reactions will be presented and discussed in the next sections.

3.2 DEFINITIONS OF TERMS

The *probability of an interaction* between species in the gas phase is determined by the effective total cross section of the interactions. Since interactions in the plasma can be of many types, such as elastic collisions, ionization, charge transfer, dissociation, and so on, each type of interaction is characterized by a specific cross section.

For a given interaction between particles of type a and particles of type b, the mean free path between interactions, λ_{ab}, is defined as the average distance that a particle of type a will traverse through a group of particles of type b before undergoing a collision and is given by

$$\lambda_{ab} = \frac{1}{\sigma_{ab} n_b} \tag{3.1}$$

where σ_{ab} = cross section for the interaction
n_b = density of particles of type b

A particle a, moving at velocity v_a, will interact with a stationary group of particles of type b at a rate given by the collision frequency ν:

$$\nu_{ab} = \frac{v_a}{\lambda_{ab}} = v_a \sigma_{ab} n_b \tag{3.2}$$

If the density of particles a is n_a, the *rate of reaction*, R, will be

$$R = n_a \nu_{ab} = n_a n_b \sigma_{ab} v_a \, (\text{cm}^3.\text{sec})^{-1} \tag{3.3}$$

The neutral, chemically unstable fragments obtained by the dissociation of the molecules are called *radicals*. Neutral single atoms, fragments of multiatomic molecules, which are unstable and very active, can also be considered radicals and will be referred to as such. In this section, atoms will be marked as A, B; molecules as M; radicals as R; and excited species, at energetic levels above the ground level, with the superscript *, for example R^*. Atomic, molecular, or radical positive ions will be marked, respectively, A^+, M^+, and R^+.

Some species are excited to electronically energetic levels at which quantum mechanical selection rules forbid spontaneous transition to ground state. As a result such species have long lifetimes and are called *metastable species*.

For some gases (e.g., molecules containing halogen or oxygen atoms), it is thermodynamically advantageous for an electron to attach itself to an atom or molecule. This type of attachment is called *electron affinity* and the gases involved are *electronegative gases*.

3.3 CHEMICAL REACTIONS

The chemical reactions in a plasma reactor can be classified into homogeneous and heterogeneous reactions. The homogeneous reactions occur between species in the gaseous phase as the result of inelastic collisions between electrons and heavy species or collisions between heavy species. Heterogeneous reactions occur between the plasma species and the solid surfaces immersed or in contact with the plasma. In the description of homogeneous reactions we shall discuss separately reactions of electrons with heavy species and reactions between heavy species.

3.3.1 Homogeneous Reactions

3.3.1.1 Reactions of Electrons with Heavy Species

The electrons in the plasma gain energy from the external electromagnetic field and transfer it to the gas to excite and sustain the plasma. The major energy transfer from the electrons to the heavy species takes place by inelastic collisions. These inelastic collisions lead to a variety of reactions, the main ones being presented in the following.

Excitation. Impact of electrons of sufficient energy with heavy targets leads to the production of excited states of atoms and molecules, as described in the reactions:

$$e + A \rightarrow e + A^{*} \tag{3.4}$$

$$e + A_2 \rightarrow A_2^{*} + e \tag{3.5}$$

$$e + AB \rightarrow e + AB^{*} \tag{3.6}$$

The *excitation* of the heavy species can be *vibrational, rotational,* or *electronic.* Atoms can only reach electronically excited states, while molecules can achieve rotationally and vibrationally excited states as well.

Electronically excited states return to ground state emitting their energy as electromagnetic radiation. This radiation accounts for the ultraviolet to visible emissions of the plasma. The relatively short lifetimes of excited species before radiative decay may preclude their participation in additional reactions. Metastable species are exceptions to this. Since metastables cannot return to their ground states by direct radiative transitions, they have long lifetimes and continue to participate in other reactions.

Dissociative Attachment. When electronegative gases are used, electrons of low energy (< 1 eV) can attach themselves to the molecules of the gas. If such attachment results in the formation of a repulsive electronically excited state, the molecule usually dissociates very fast ($\approx 10^{-13}$ sec), producing a negative ion according to equation

$$e + AB \rightarrow A + B^{-} \tag{3.7}$$

Negative ions can also be produced by *dissociative ionization* reactions:

$$e + A_2 \rightarrow A^{+} + A^{-} + e \tag{3.8a}$$

$$e + AB \rightarrow A^{+} + B^{-} + e \tag{3.8b}$$

The reaction described by Eq. (3.7) is also called *dissociative capture,* while the reactions described by Eq. (3.8a) and Eq. (3.8b) are also called *ion-pair formation* reactions.

The threshold of the reaction described by Eq. (3.8b) is usually high compared to that of dissociative capture described by Eq. (3.7) because the formation of positive ions requires considerable energy. The maximum cross section is observed between 20 eV and 30 eV [6].

Some examples of dissociative attachment reaction are

$$e + O_2 \rightarrow 2O + e \rightarrow O + O^- \tag{3.9}$$

$$e + CF_4 \rightarrow CF_3 + F + e \rightarrow CF_3 + F^- \tag{3.10}$$

The reaction described in Eq. (3.9) is important in oxygen plasmas used for the stripping of photoresist in microelectronic processing. The reaction described in Eq. (3.10) is important in plasma etching of Si, SiO_2, and Si_3N_4. Atomic oxygen reacts easily with the organic materials used for photoresists to form volatile CO, CO_2, and H_2O, while atomic fluorine is the active species for etching silicon and silicon compounds in CF_4 discharges [7]. Halogens usually dissociate in a cold plasma according to Eq. (3.7) and Eq. (3.8b). For halides, which dissociate in cold plasmas according to Eq. (3.7) the electron energy required ranges from 1.88 eV for HF to 0.03 eV for HI [8].

The cross sections of the dissociative attachment reactions are dependent on the kinetic energy of the electrons and on the properties of the molecules involved. In certain cases the dependence of the cross section on the kinetic energy has the shape of a narrow resonance peak [6].

Typical values of cross sections for formation of negative ions by dissociative capture are given in Table 3-1 together with the energy of the impacting electron, W_{max}, at the maximum cross section. The examples of cross sections span a range from relatively high values of 10^{-14} cm^{-2}, at very low electron energies for I^{-1}, to very low values of 10^{-21} for H^-, at medium to high energies.

Dissociation. An inelastic collision of an electron with a molecule can cause its dissociation without the formation of ions, according to

$$e + A_2 \rightarrow 2A + e \tag{3.11}$$

$$e + AB \rightarrow e + A + B \tag{3.12}$$

TABLE 3-1 Cross Section for Dissociative Capture (adapted after [6], reprinted by permission of John Wiley & Sons, Inc.)

Molecule	Ion	W_{max} (eV)	σ (W_{max}) (10^{-17} cm^2)
HI	I^-	~0.0	2300
I_2	I^-	0.3	300
HBr	Br^-	0.28	27
HCl	Cl^-	0.81	1.99
O_2	O^-	6.7	0.143
CO_2	O^-	8.03	0.0482
H_2O	H^-	8.6	0.13
H_2	H^-	3.75	0.000016

The dissociation of molecules occurs only through their vibrational and electronic excitations. Most dissociation of molecules by slow electrons is induced by electronic excitation. The dissociation will occur only if the molecule is excited above a threshold value. For example, for H_2 the threshold value is 8.8 ± 0.2 eV and 9.6 ± 0.5 eV for formation of N atoms (see page 72 of [4]). The maximum cross section for electron impact dissociation of hydrogen molecules was observed at an electron energy of 16.5 eV, and its value is 9×10^{-17} cm^2 (see page 72 of [4]). A molecule in a vibrational excited state can be also further excited to dissociative states even by low-energy electron impact.

The cross section for vibrational energy excitation by electron impact can vary by two orders of magnitude from one molecule to another. Certain molecules, such as N_2, CO, and CO_2, have large cross sections of the order of 1-5 $\times 10^{-16}$ cm^2, while others, such as H_2 and O_2 have cross sections that are smaller by a factor of 10–50 [9]. The dissociation of the excited molecule into neutral fragments occurs in a very short time of about 10^{-13} sec. Because the lifetime for radiative decay is longer than 10^{-7}–10^{-8} sec [4], the probability is almost 100% that the molecule will return to a lower energy state by dissociation and not by radiative decay.

Table 3-2 presents values of cross sections for the formation of neutrals and positive ions from the dissociation of CH_4 by collision with fast 100 eV electrons. The values for the formation of neutral radicals from CH_4 span the range of 0.1–2.4 $\times 10^{-16}$ cm^2.

TABLE 3-2 Cross Sections for Formation of Neutral and Positively Charged Fragments from CH_4 (from [6], reprinted by permission of John Wiley & Sons, Inc.)

Positive Ions		Neutral Fragments	
Ion	σ ($\times 10^{16}$), cm^2	Species	σ ($\times 10^{16}$), cm^2
H^+	0.04	H	2.4
H_2^+	0.02	H_2	0.8
C^+	0.05	C	—
CH^+	0.14	CH	0.1
CH_2^+	0.28	CH_2	0.2
CH_3^+	1.5	CH_3	1.2
CH_4^+	1.8		—

The *dissociative attachment, dissociative ionization,* and *dissociation* reactions are a major source for the production of atoms, free radicals and negative ions in the cold plasma.

Ionization. Ionization in discharges taking place in a molecular gas occurs predominantly by electron impact that can produce positive or negative, atomic, or molecular ions:

$$e + A_2 \rightarrow A_2^+ + 2e \tag{3.13a}$$

$$e + A_2 \rightarrow A_2^- \tag{3.13b}$$

$$e + A_2 \rightarrow A^+ + A + 2e \tag{3.13c}$$

$$e + AB \rightarrow 2e + A^+ + B \tag{3.13d}$$

Ionization potentials of some atoms and molecules are shown in Table 3-3. As can be seen in the table, the ionization potentials span the range from 8 to 25 eV, which are well above the mean electron energy in a cold plasma. Thus, only

TABLE 3-3 Ionization Potentials of Atoms and Molecules (in electron volts)

Neutral	Ion	Ionization Potential
Ar	Ar^+	15.8
Ar^+	Ar^{++}	27.6
F	F^+	17.4
H	H^+	13.6
He	He^+	24.6
N	N^+	14.5
O	O^+	13.6
Si	Si^+	8.1
CH_4	CH_4^+	13
C_2H_2	$C_2H_2^+$	11.4
H_2	H_2^+	15.4
HF	HF^+	17
H_2O	H_2O^+	12.6
N_2	N_2^+	15.6
O_2	O_2^+	12.2
SiH_4	SiH_4^+	12.2

electrons in the high-energy tail of the electron distribution can contribute to the ionization reactions. Ionization may sometimes occur also through a stepwise process, in which impact by lower-energy electrons cause first the excitation of the molecule to a metastable state. Subsequent impact of the metastable with an electron can cause its ionization. However, such a mechanism is not a dominant one.

The ionization reaction described in Eq. (3.13b) is also called a *resonance* or *nondissociative capture* reaction and can be followed by a dissociation, according to Eq. (3.7), as illustrated for oxygen:

$$e + O_2 \rightarrow O_2^- \rightarrow O + O^- \tag{3.14}$$

The threshold potential for this reaction is 4.53 eV, as compared to 17.28 eV for the direct formation of the $O^+ + O^-$ pair [8].

The reaction described in Eq. (3.13d) can produce a large number of different ions, when molecules containing a large number of atoms, such as higher hydrocarbons, are involved. An especially large number of ions is obtained from benzene, ranging from $C_6H_6^+$ to H^+ and covering the whole range $C_xH_y^+$ with $0 \leq x,\ y \leq 6$ [4].

Values of cross sections for the formation of different ionized fragments of CH_4 are shown in Table 3-2. The values span a two orders of magnitude range from 0.04 to 1.8×10^{-16} cm^2.

The dependence of the cross section of ionization for several molecules in the ground state on the electron energy is illustrated in Fig. 3-1.

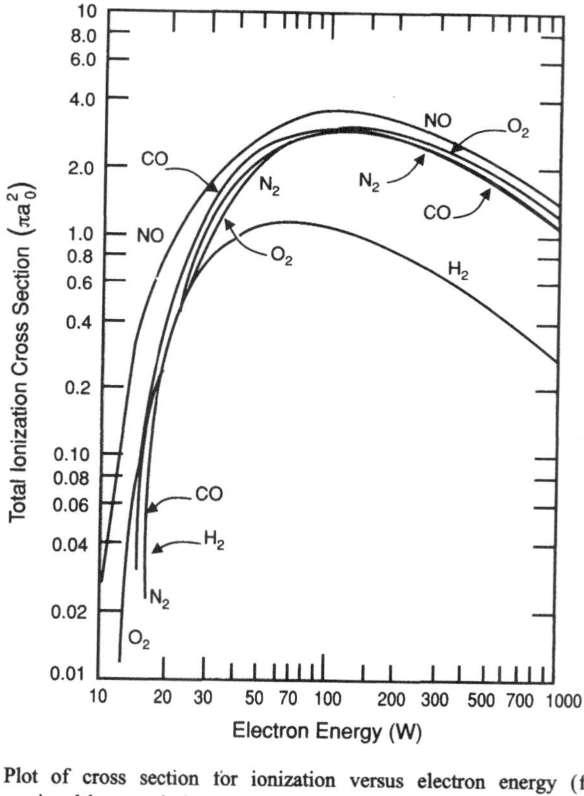

Fig. 3-1 Plot of cross section for ionization versus electron energy (from [6], reprinted by permission of John Wiley & Sons, Inc.).

Positive ions are usually formed in most ionization processes prevalent in the cold plasmas. Nevertheless, when the atom or molecule involved in the reaction possess electron affinity, negative ions can also form in the cold plasmas, according to Eqs. (3.7), (3.8a), and (3.13b). The process described by Eq. (3.13b) is a radiative attachment that is very slow and its typical cross section is 10^{-19} cm^2 at a few electron volts [10].

Oxygen, halogens, and halogen-containing molecules are some of the molecules with a tendency to form negative ions. For these molecules the cross sections have high values at low electron energies, as illustrated in Table 3-1. In addition to direct ionization, or dissociative attachment described earlier, negative ions can also form through radiative or nondissociative electron capture (e.g., I^-, Br^-).

Recombination. The charged particles (electrons and ions) are lost from the plasma by recombination of particles of opposite charges. Recombinations that take place between electrons and atomic ions are accompanied by emission of electromagnetic radiation and are called *radiative recombinations*:

$$e + Ar^+ \rightarrow Ar + hv \tag{3.15}$$

where h = Planck's constant
v = radiation frequency

The term hv indicates release of radiation energy.

On the other hand, the release of energy during the recombinations of electrons with molecular ions can cause the dissociation of the molecule by a *dissociative recombination* reaction:

$$AB^+ + e \Leftrightarrow AB^* \rightarrow A^* + B \tag{3.16a}$$

$$e + A_2^+ \rightarrow 2A \tag{3.16b}$$

The rates of recombination of electrons with atomic ions, Eq. (3.15), are very low, in the range of 10^{-13} $cm^3.sec^{-1}$. The rates of recombination with molecular ions, Eqs. (3.16a) and (3.16b), are much higher, in the range 10^{-9} to 10^{-10} $cm^3.sec^{-1}$ [8].

3.3.1.2 Reactions Between Heavy Species

Reactions between heavy species are those occurring during collisions of molecules, atoms, radicals, and ions. The reactions between the heavy species can be grouped in two subcategories, namely, *ion-molecule* and *radical-molecule reactions*. Ion-molecule reactions are those that involve at least one ion. Radical-molecule reactions are the reactions occurring between neutral species only.

The density of molecules in a gas at a pressure of 1 torr is $\sim 3.5 \times 10^{16}$ cm^{-3}. In a plasma with a mean electron energy of 1 eV, sustained in a gas characterized by a molecular dissociation energy of about 5 eV and an ionization energy of 10 eV, the density of dissociated molecules can be estimated to be 10^{14} cm^{-3} [11]. The density of ions and electrons in a cold plasma is usually about 10^9 to 10^{11} cm^{-3}. Thus, in a cold plasma, the three densities are related by

$$n_i \ll n_r \ll n_n \tag{3.17}$$

where n_r is the radical density.

Because of these relative particle densities, it can be expected that the radical-molecule reactions will be generally more important than will ion-molecule reactions in plasma chemistry. This is indeed the case in many processes. However, ion-molecule reactions can have significant contributions to plasma chemistry and can enhance the rate of the total reaction occurring in the plasma

[12]. Following is a description of the more important types of plasma reactions in each subcategory.

Ion-Molecule Reactions

Recombination of ions. Two colliding ions can recombine to form a molecule in the ground state and release their energy trough emission of radiation:

$$A^+ + B^- \rightarrow AB + h\nu \tag{3.18}$$

A collision between two ions can also result in the *neutralization* of the ions by formation of two excited atoms:

$$A^+ + B^- \rightarrow A^* + B^* + h\nu \tag{3.19}$$

Because the total energy of the recombining ions is generally larger then the total energy of the excited neutrals the surplus energy is released through radiation.

Ion-ion recombination can also take place trough a three-body collision:

$$M + A^+ + B^- \rightarrow AB + M \tag{3.19a}$$

The two-body reaction described by Eq. (3.18) is important at very low pressures, while three-body recombination occurs at pressures above 0.1 mtorr.

Charge transfer. An electric charge may be transferred during a collision between an ion and a neutral particle. The transfer can take place between identical partners, for example,

$$A + A^+ \rightarrow A^+ + A \tag{3.20}$$

or dissimilar partners:

$$B_2 + A^+ \rightarrow B_2^+ + A \tag{3.21a}$$

$$A^+ + BC \rightarrow A + BC^+ \tag{3.21b}$$

Although it appears from Eq. (3.20) that this reaction has no effect, it results in fact in a slower ion and a faster neutral. The symmetric processes described by Eq. (3.20) are characterized by large cross sections. At low energies, the asymmetric processes involving molecules — (described by Eq. (3.21)) — are frequently characterized by large cross sections, whereas the ones involving atoms are characterized by small cross sections. Charge transfer can occur in reactions with either positive ions as just described or with negative ions.

When the charge transfer takes place with dissociation of the colliding species, according to Eq. (3.22), the reaction is called *charge transfer with dissociation*:

$$A^+ + BC \rightarrow A + B^+ + C \tag{3.22}$$

The reaction rate constants of ion-molecule charge transfer reactions are typically of order of 10^{-10}–10^{-12} $cm^3.sec^{-1}$ [13].

Transfer of heavy reactants. This type of ion-molecule reaction results in the formation of new compound species as in Eq. (3.23):

$$A^+ + BC \rightarrow \begin{cases} AB^+ + C \\ AB + C^+ \end{cases} \tag{3.23}$$

The reaction is sometimes also called *ionization by interchange*. Examples of such transfer reactions are:

$$N^+ + O_2 \rightarrow NO^+ + O \tag{3.24}$$

$$N + O_2^+ \rightarrow NO^+ + O \tag{3.25}$$

These reactions of heavy reactant transfer occur with rate coefficients greater than 10^{-9} $cm^3.sec^{-1}$. Detailed discussions of ion-molecule reactions can be found in [14].

Associative detachment. In a collision between a negative ion and a radical, the ion can attach itself to the radical, neutralizes by releasing the electron, and forms a new compound. This is an *associative detachment* reaction described by the equation

$$A^- + BC \rightarrow ABC + e \tag{3.26}$$

Radical-Molecule Reactions

Radical-molecule reactions are those where only neutral species act as reactants. The active radicals may be either multiatom radicals or single atoms, fragments of multiatomic molecules. The radicals are unstable and chemically very active. Representative radical-molecule reactions are presented in the following.

Electron transfer. This is a reaction between two neutral particles resulting in the formation of two ions by the transfer of an electron between the colliding neutrals:

$$A + B \rightarrow A^+ + B^- \tag{3.27}$$

This type of reaction requires that at least one molecule has very high kinetic energy and its occurrence in cold plasmas is therefore rare.

Ionization. A collision between two energetic neutrals can cause the ionization of one of them:

$$A + B \rightarrow A^+ + B + e \tag{3.28}$$

Penning ionization/dissociation. The Penning reactions occur during collisions involving energetic metastable species. In the collisions of metastables (B^*) with neutral species, the excited metastables transfer their excess energy to the target, causing ionization or dissociation according to Eq. (3.29a) or Eq. (3.29b):

$$B^* + A \rightarrow A^+ + B + e \tag{3.29a}$$

$$B^* + A_2 \rightarrow 2A + B \tag{3.29b}$$

The Penning processes are particularly important in plasmas sustained in mixtures, which include gases such as argon and helium, that are characterized by several long-lifetime metastable states. Furthermore, Penning ionization has a large cross section, which enhances the probability of this process.

For Penning reactions to take place, the energy of the metastable species must be higher than the ionization or dissociation potential of the other species participating in the reaction. Metastable energies are in the range of 0–20 eV (11.5 eV for Ar, 16.6 eV for Ne). As illustrated in Fig. 3-2 for nitrogen, for certain gases at lower electron energies, the cross sections for metastable excitation are much higher than the cross sections for ionization. In such conditions, electron impact causes formation of a larger number of metastable particles than ions. The metastable species can accumulate energy and reach energy levels higher than the ionization threshold of the interacting species, and this can release their energy by Penning ionization. Penning dissociation is a similar process that results in the dissociation of the target molecule into neutral radicals instead of its ionization.

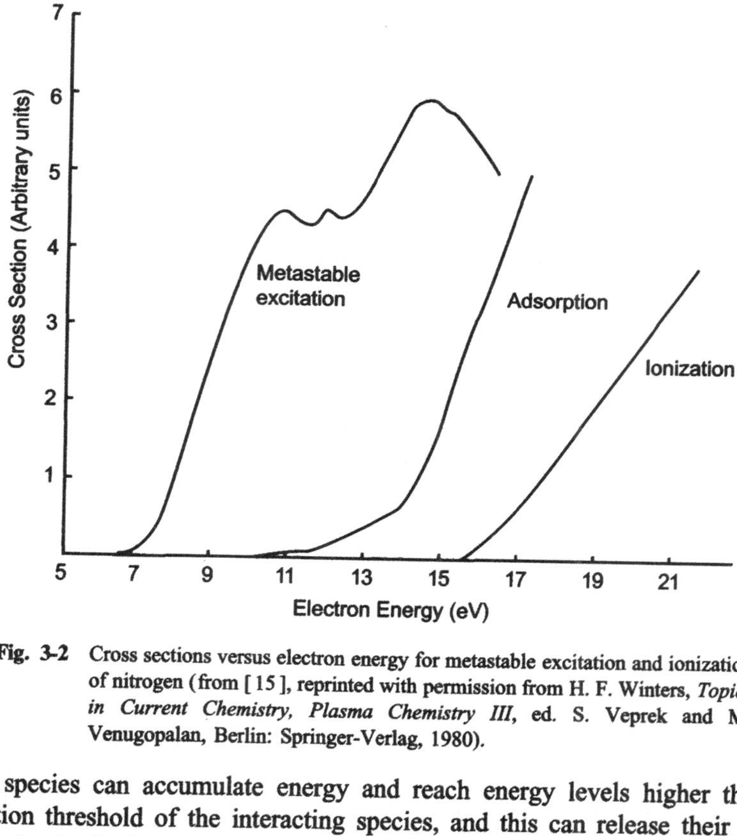

Fig. 3-2 Cross sections versus electron energy for metastable excitation and ionization of nitrogen (from [15], reprinted with permission from H. F. Winters, *Topics in Current Chemistry, Plasma Chemistry III*, ed. S. Veprek and M. Venugopalan, Berlin: Springer-Verlag, 1980).

During collisions of metastable atoms with certain molecules (for example, H_2O, NH_3, SF_6), the metastable atom can release its energy by becoming itself ionized according to the reaction:

$$B^* + M \rightarrow B^+ + M + e \tag{3.30}$$

When the molecule involved in the reaction is SF_6, a negative ion is formed through the attachment of the released electron to the SF_6 molecule:

$$B^* + SF_6 \rightarrow B^+ + SF_6^- \tag{3.31}$$

The cross section for the last reaction is $(1.7 \pm 0.11) \times 10^{-12}$ cm^2.

Attachment of atoms. These are reactions similar to the associative detachment reactions of the ion-molecule type described by Eq. (3.26) but involve only neutral species:

$$A + BC + M \rightarrow ABC + M \tag{3.32}$$

Disproportionation. The disproportionation reaction ressembles to the ion-molecule transfer of heavy reactants reaction, but it occurs between neutral species according to

$$A + BC \rightarrow AB + C \tag{3.33}$$

Recombination of radicals. Collisions between the chemically active radicals can cause their recombination into stable molecules. The requirement of simultaneous conservation of both the energy and momentum prevents the direct recombination of two monoatomic radicals. Recombination of monoatomic radicals can therefore take place only through a multibody collision, when the third body might be another particle in the plasma or a solid surface in contact with the plasma. The large number of degree of freedom of multiatomic free radicals allows the redistribution of internal energy, and the conservation of momentum and energy can be achieved in a two-body recombination. Therefore, the collisional efficiency of recombination for multiatomic radicals becomes close to unity.

Radicals usually combine through two type of reactions:

- *Disproportionation*

$$2C_2H_5 \rightarrow C_2H_4 + C_2H_6 \tag{3.34}$$

- *Combination*

$$2C_2H_5 \rightarrow C_4H_{10} \tag{3.35}$$

Chemiluminescence. Excitation of an atom or a molecule can take place in the plasma during a collision with another atom. The excitation can occur during a chemical reaction (e.g., disproportionation or recombination), but can also take place without the occurrence of a chemical reaction. The two possibilities are illustrated by Eq. (3.36) and Eq. (3.37):

$$A^* + BC \rightarrow A + BC^* \tag{3.36}$$

$$B + CA \rightarrow BC^* + A \tag{3.37}$$

Equation (3.36) describes a Penning-type reaction. The excited radical BC^* usually returns from its excited state to its energetic ground level by radiative decays:

$$AB^* \rightarrow AB + h\nu \tag{3.38}$$

Equations (3.36)–(3.38) describe the chemiluminescence reactions that can occur in the plasma and contribute to its luminescence [16].

3.3.1.3 Comments on Homogeneous Reactions

At pressures of about 1 torr, the free radical concentration in a cold plasma is up to 10^5 higher than the ion concentration. Therefore radical-molecule reactions usually occur at faster rates than do ion-molecule reactions. It was found experimentally that addition of radical forming halogen compounds to hydrocarbon monomers accelerates significantly the polymerization rate and reduces the hydrogen to carbon ratio in the polymer [17]. However, in certain conditions, superposition of ion-molecule reactions can strongly accelerate the overall reaction kinetics. An example of such a behavior was observed in the deposition of silicon from plasmas of silicon tetrachloride ($SiCl_4$), hydrogen, and argon.

In plasmas of $SiCl_4 + H_2$ the dissociation of $SiCl_4$ and deposition of Si occur through radical-molecule reaction [12]. In plasmas of $SiCl_4$ + Ar the reactions occur through ion-molecule processes and the overall reaction rates are much smaller than in the $SiCl_4 + H_2$ plasma. However, the superposition of ion-molecule reactions on radical-molecule reactions in plasmas of $SiCl_4 + H_2$ + Ar causes a strong enhancement in the overall reaction rate and the Si deposition rate. In this case the enhancing effect of the Ar was explained by an increase in the H radical concentration, through interactions of Ar^+ and metastable Ar^* with hydrogen molecules [12].

The homogeneous chemical reactions occur predominantly in the cathode fall and positive column regions of the glow discharge. The highest rate of reaction is observed in the cathode fall, where the strength of the electric field is highest (see Fig. 2-4 in Sec. 2.1.2). However, in situations where the cathode fall occupies only a small fraction of the total reaction volume, the overall reaction rate can be significantly affected also by the positive column.

The occurrence of a specific reaction is obviously a function of the gases involved and is determined by the minimal energy required for that reaction. If the plasma species involved have enough energy for the reaction to occur, the reaction and its rate are also determined by the cross section of the reaction.

The cross section is also a function of the energy of the species, as already indicated in Fig. 3-1 and Fig. 3-2. The cross sections and reaction rate constants for the various reactions cover a wide range of values, even for a plasma sustained in a simple molecular gas. This is illustrated in Table 3-4 for an oxygen plasma.

As can be seen in Table 3-4, the values of the cross sections for reactions occurring in an oxygen plasma span a range of values from 10^{-20} to 10^{-16} cm^2. Even for the same type of reaction, the cross sections can vary up to two orders of magnitude from one molecule to another. Typical cross sections for various types of reactions just described can be found in [4, 5, 10, and 18].

A characteristic situation encountered in cold plasmas is illustrated in Fig. 3-3, which shows the cross sections for electron-impact ionization of argon together with the Druyvesteyn energy distribution of the electrons in an argon plasma, at a typical electron temperature T_e = 5 eV. As illustrated in Fig. 3-3, the majority of the electrons in this plasma have energies that are either smaller than the threshold for ionization or correspond to very low ionization cross sections.

TABLE 3-4 Cross Sections and Reaction Rate Constants in an Oxygen Plasma (from [5], reprinted by permission of John Wiley & Sons, Inc.)

Reaction	k	σ_{max} (cm^2)
Ionization		
1. $e + O_2 \rightarrow O_2^+ + 2e$		2.72×10^{-16}
2. $e + O \rightarrow O^+ + 2e$		1.54×10^{-18}
Dissociative ionization		
3. $e + O_2 \rightarrow O^+ + O$		1.0×10^{-16}
Dissociative attachment		
4. $e + O_2 \rightarrow O^- + O$		1.41×10^{-18}
5. $e + O_2 \rightarrow O^- + O + e$		4.85×10^{-19}
Dissociation		
6. $e + O_2 \rightarrow 2\,O + e$		2.25×10^{-18}
Metastable formation		
7. $e + O_2 \rightarrow O_2(^1\Delta_g) + e$		3.0×10^{-20}
Charge transfer		
8. $O^+ + O_2 \rightarrow O_2^+ + O$	2×10^{-11} cm^3/sec	
9. $O_2^+ + O \rightarrow O^+ + O_2$		8×10^{-16}
10. $O_2^+ + O_2 \rightarrow O_3^+ + O$		1×10^{-16}
11. $O_2^+ + 2\,O_2 \rightarrow O_4^+ + O_2$	2.8×10^{-30} cm^6/sec	
12. $O^- + O_2 \rightarrow O_2^- + O$	2.5×10^{-14} cm^3/sec at $E/p = 20$ V/cm torr; 3.4×10^{-12} cm^3/sec at $E/p = 45$ V/cm torr	
13. $O^- + O_3 \rightarrow O_3^- + O$	5.3×10^{-10} cm^3/sec	
14. $O^- + 2\,O_2 \rightarrow O_3^- + O_2$	$1.0 \pm 0.2 \times 10^{-30}$ cm^6/sec	
15. $O_2^- + O \rightarrow O^- + O_2$	5×10^{-10} cm^3/sec	
16. $O_2^- + O_2 \rightarrow O_3^- + O$		$< 10^{-18}$
17. $O_2^- + O_3 \rightarrow O_3^- + O_2$	4.0×10^{-10} cm^3/sec	
18. $O_2^- + 2\,O_2 \rightarrow O_4^- + O_2$	3×10^{-31} cm^6/sec	
19. $O_3^- + O_2 \rightarrow O_2^- + O_3$		4×10^{-17}
20. $O_4^- + O \rightarrow O_3^- + O_2$	4×10^{-10} cm^3/sec	
21. $O_4^- + O_2 \rightarrow O_2^- + 2\,O_2$	6×10^{-15} cm^3/sec	
Detachment		
22. $O^- + O \rightarrow O_2 + e$	3.0×10^{-10} cm^3/sec	
23. $O^- + O_2 \rightarrow O + O_2 + e$		7×10^{-16}
24. $O^- + O_2(^1\Delta_g) \rightarrow O_3 + e$	$\sim 3 \times 10^{-10}$ cm^3/sec	
25. $O_2^- + O \rightarrow O_3 + e$	5.0×10^{-10} cm^3/sec	
26. $O_2^- + O_2 \rightarrow 2\,O_2 + e$		7×10^{-16}
27. $O_2^- + O_2(^1\Delta_g) \rightarrow 2\,O_2 + e$	$\sim 2 \times 10^{-10}$ cm^3/sec	
Electron-ion recombination		
28. $e + \left\{\begin{matrix} O \\ O_2^+ \\ O_3^+ \\ O_4^+ \end{matrix}\right\} \rightarrow \left\{\begin{matrix} O \\ 2\,O \\ O + O_2 \\ 2\,O_2 \end{matrix}\right\}$	$\lesssim 10^{-7}$ cm^3/sec	

TABLE 3-4 *cont.* Cross Sections and Reaction Rate Constants in an Oxygen Plasma (from [5], reprinted by permission of John Wiley & Sons, Inc.)

Reaction	k	σ_{max} (cm^2)
Ion-ion recombination		
29. $\left\{\begin{matrix} O^- \\ O_2^- \\ O_3^- \\ O_4^- \end{matrix}\right\} + \left\{\begin{matrix} O^+ \\ O_2^+ \\ O_3^+ \\ O_4^+ \end{matrix}\right\} \rightarrow \left\{\begin{matrix} O \\ O_2 \end{matrix}\right\}$	$\sim 10^{-7}$ cm^3/sec	
Atom recombination		
30. $2\,O + O_2 \rightarrow 2\,O_2$	2.3×10^{-33} cm^6/sec	
31. $3\,O \rightarrow O + O_2$	1.5×10^{-34} cm^6/sec	
32. $O + 2\,O_2 \rightarrow O_2 + O_2$	1.9×10^{-35} exp $(2100/RT)$$cm^6$/sec	
33. $O + O_3 \rightarrow 2\,O_2$	2.0×10^{-11} exp $(-4790/RT)$$cm^3$/sec	
34. $O \xrightarrow{\text{wall}} O_2$	$\gamma = 1.6 \times 10^{-4}$ to 1.4×10^{-2} ($T = 20 - 600$ °C)	

As a result, only the small fraction of electrons that have energies in the high energy tail of the distribution curve will cause the ionization of the gas.

In most ionization processes prevalent in cold plasmas, positive ions are formed. However, when the atom or molecule involved in the reaction possess electron affinity, negative ions can also form according to Eqs. (3.8a), (3.7), and (3.13b). Electron attachment to neutral atoms happens most easily to those atoms that have almost filled outer electron shells, for example, oxygen, halogens, or halogen-containing molecules. The source of the electron can be another atom, molecule, or a free electron.

The negative ions formed by the nondissociative capture of electrons by complex molecules according to Eq. (3.13b) are long-lived negative ions. The long lifetimes of such ions are due to the fast distribution of the electronic energy over vibrational degrees of freedom, thus making difficult the reverse transition necessary for detachment of electrons. The coefficient for attachment of electrons to SF_6 molecules by this mechanism was found to be 1.24×10^9 $cm^{-1}.torr^{-1}$, and the lifetime of an excited SF_6^- ion was estimated as 25 μs [19].

The formation of negative ions causes the replacement of highly mobile electrons with much slower heavier ions. Large densities of negative ions will change the ambipolar electron diffusion coefficient and thereby the fluxes of electrons and positive ions. This results in the reduction of the electrical conductivity of the plasma that may cause plasma instabilities. Such instabilities, in turn, can impede the control of the process and make it difficult to achieve uniformity of results.

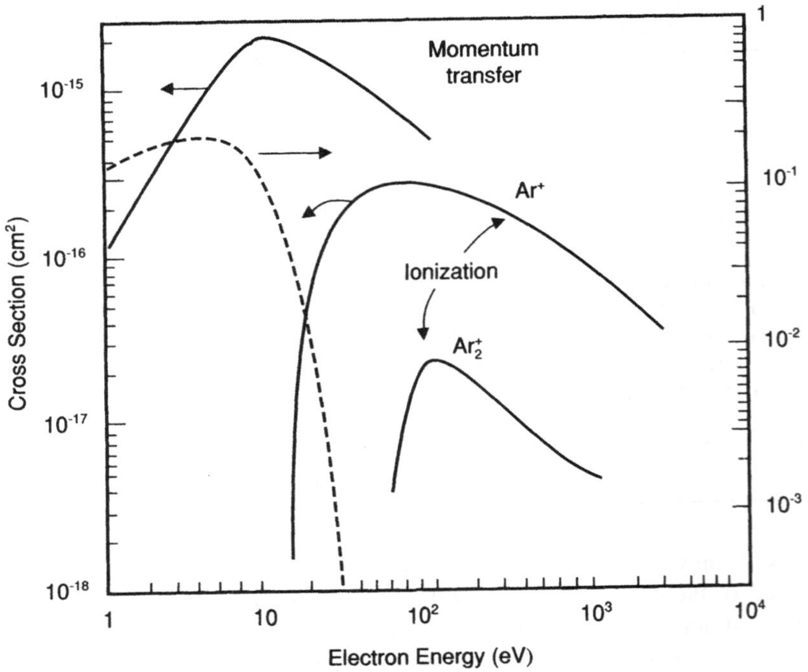

Fig. 3-3 Cross sections versus electron energy (—) and Druyvesteyn energy distribution for $T_e = 5$ eV (- -).

3.3.2 Heterogeneous Reactions

The heterogeneous reactions are those which occur as a result of interactions between a solid surface (S) exposed to the plasma and plasma species. The plasma species can be an individual atom (A, B), a monomer molecule (M), a simple radical (R), or a polymer (P) formed in the plasma. Typical heterogeneous reactions are presented next.

Adsorption. When molecules, monomers, or radicals from the plasma come in contact with a surface exposed to the plasma they can be adsorbed on the surface. The adsorption reaction can be described by

$$M_g + S \rightarrow M_s \tag{3.39a}$$

$$R_g + S \rightarrow R_s \tag{3.39b}$$

The indexes g and s indicate, respectively, a specie in the gas or solid phase. Most radicals interact with surfaces, and consequently the composition of the deposited film will be largely determined by the relative fluxes of all film-forming species [20].

Recombination or compound formation. Atoms or radicals from the plasma can react with species already adsorbed on the surface to combine and form compounds according to

$$S - A + A \rightarrow S + A_2 \tag{3.40a}$$

$$S - R + R_1 \rightarrow S + M \tag{3.40b}$$

$S - A$ indicates an atom A adsorbed on the surface S.

During recombination, the energy of the particles participating in the reaction is usually released as heat to the surface. The rate of surface recombination is strongly dependent on the catalytic properties of the surface [21]. For example, the probability of recombination of fluorine atoms ($F + F \rightarrow F_2$) on Teflon is small (7×10^{-5}) [22], while on copper it is approximately three orders of magnitude higher [23]. Surfaces that are good catalysts for recombinations can be passivated by the addition of small amounts of impurities to the gas.

Metastable deexcitation. An excited metastable specie M^* from the plasma can release its energy and return to the ground state by collision with a solid surface. The reaction

$$S + M^* \rightarrow S + M \tag{3.41}$$

describes the metastable deexcitation reaction.

Sputtering. A surface exposed to the plasma is always negative relative to the plasma causing positive ions from the plasma to accelerate toward the surface. If an ion A^+ arrives at the surface with sufficient energy, it can remove an atom from the surface:

$$S - B + A^+ \rightarrow S^+ + B + A \tag{3.42}$$

This process is called sputtering. The atom B in Eq. (3.42) can be in this case either an atom from the solid surface or an atom adsorbed on it. The sputtered neutral atom enters the plasma with a kinetic energy of several electron volts.

Almost 95% of the sputtered atoms originate within a few angstroms of the surface as a result of a collision cascade within the solid.

Polymerization. Radicals in the plasma can react with radicals adsorbed on the surfaces and form polymers:

$$R_g + R_s \rightarrow P_s \tag{3.43a}$$

$$M_g + R_s \rightarrow P'_s \tag{3.43b}$$

Polymerization or radical formation can also happen between two species adsorbed on the solid surface:

$$R_s + R'_s \rightarrow P_s \tag{3.44a}$$

$$M_s + R_s \rightarrow R'_s \tag{3.44b}$$

3.3.2.1 General Remarks

The recombination reactions described by Eq. (3.40a) and Eq. (3.40b) can cause either the deposition of a solid film or etching of the surface if the formed compound is volatile. The polymerization reactions, Eq. (3.43a) to Eq. (3.44a), also result in a deposition of a film.

In metastable deexcitation, Eq. (3.41), or in recombination, Eq. (3.40b), reaction atoms on the surface act as a third body.

When a molecular ion collides with a solid surface, it is generally dissociated into its constituent atoms. Upon collision, part of the kinetic energy of the ion is transferred into internal energy, which is generally much larger than the energy required to break the chemical bond. This causes the dissociation of the molecular ion on impact on the solid surface. For example, the energy required to break the bond between the atoms of the nitrogen molecule is about 9 eV. It was found that the probability of breaking up an N_2^+ ion on collision with a solid surface has a threshold of around 9 eV and reaches 100% at ~ 100 eV, as shown in Fig. 3-4 [15].

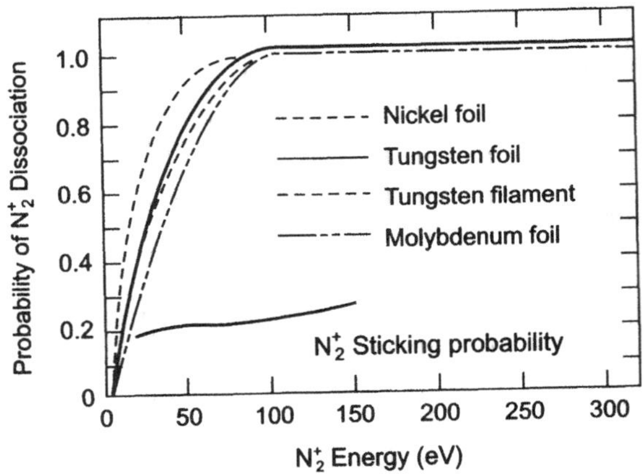

Fig. 3-4 Probability that an N_2^+ will dissociate upon collision with a surface versus ion energy (from [15], reprinted with permission from S. Veprek and M. Venugopalan, eds., *Topics in Current Chemistry, Plasma Chemistry III*, Berlin: Springer-Verlag, 1980).

These energies correspond to the substrate biases or self-biases encountered in a DC- or RF-excited plasma. Some of the atoms or radicals produced by the dissociation of molecular ions are reflected from the surface, while others are incorporated into its crystal lattice. The reflected atoms are likely to react with the surrounding surfaces or may be incorporated in the growing film. It was found, for example, that there is a 20% probability that the nitrogen atoms resulting from the dissociation of the N_2^+ ion will stick to the surface and eventually form a thin nitride layer of 20–30 Å. The other 80% of the nitrogen atoms are reflected into the gas phase. Similarly, bombardment of a surface with O_2^+, or CH_4^+ will cause formation of a surface layer of oxide or carbide.

Negative ions are usually ignored in the mechanistic discussions of plasma processing of surfaces. Because any surface in contact with a plasma is at a negative potential relative to the plasma, the negative ions created in the plasma cannot reach the surfaces and are usually ignored in the discussion of plasma interaction with surfaces. However, in some situations negative ions will participate in surface reactions. Such situations arise whenever a negatively biased surface has a tendency to form negative secondary ions when bombarded with positive ions. These negative ions will be accelerated into the plasma by the sheath potential, and if the operating pressure and interelectode distance are sufficiently low, the negative ions can traverse the plasma arriving at the opposite electrode with substantial kinetic energy. This effect has been observed during sputter deposition of halides and plasma deposition of high-temperature superconducting materials [20].

3.4 THE CHEMICAL REACTION CHAIN

The chemical processes in the plasma follow several steps: *initiation, propagation, termination*, and *reinitiation* [12, 17].

In the initiation stage free radicals or atoms are produced by collision of energetic electrons or ions with molecules. The radical formation takes place by dissociation of molecules in the gaseous phase or of molecules adsorbed on the surface of the substrate or on the deposited film. Both molecules and radicals are adsorbed on the surfaces exposed to the plasma.

The propagation step of the reactions can also take place in both the gas phase and on the modified surface (deposited film, etched surface). In the gas phase, propagation involves the interaction between radicals, ions, and molecules in ion-molecule and radical-molecule reactions. Propagation takes place on the solid surface through interactions of surface free radicals with either gas phase or adsorbed molecules, radicals, or ions.

In the termination step, reactions similar to those described for propagation result in the formation of the final product.

Reinitiation develops when radicals, which are formed through conversion of the deposited film or dissociation of polymers, again enter the reaction chain. The conversion to radicals can be caused by impact of energetic particles or by photon absorption.

The complexity of the reaction chain is illustrated in Fig. 3-5 for PECVD deposition of amorphous silicon from $Ar + H_2 + SiCl_4$ plasmas [24]. The initiation, propagation, and termination steps are shown with the various types of reactions taking place in each step.

During PECVD deposition processing it is important to prevent the termination of the main reaction chain in the homogeneous gas phase or the occurrence of homogeneous reactions that can result in formation of solid powders. If not prevented, these homogeneous reactions will affect the film deposition rate and the quality of the deposited films. They can also cause contamination, whose prevention is becoming extremely and increasingly important in the microelectronics industry.

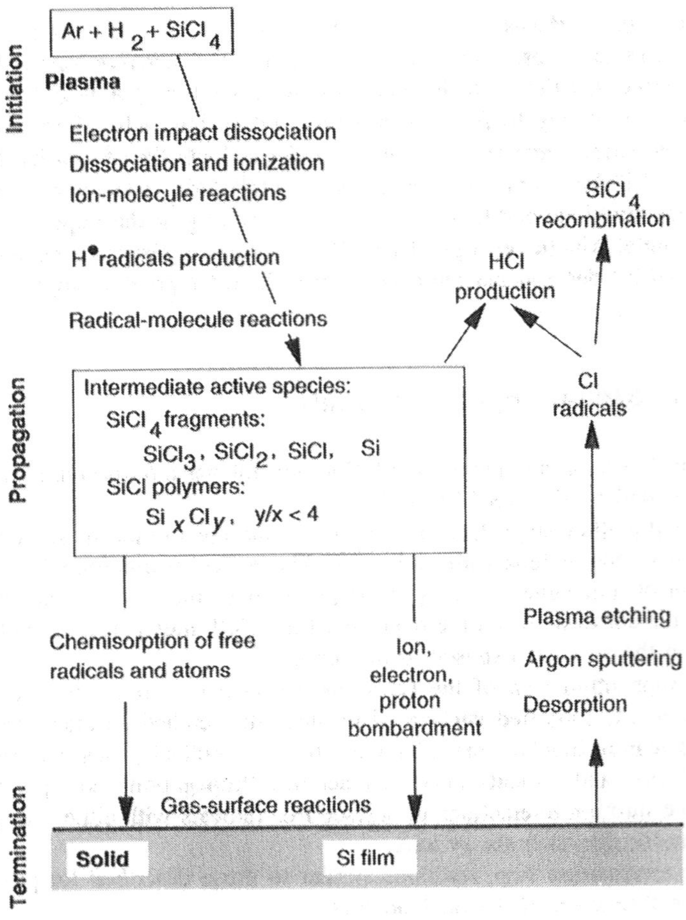

Fig. 3-5 The reaction chain during deposition of silicon from Ar + H_2 + $SiCl_4$.

A schematic summary of the different types of reactions that can take place in a cold plasma reactor is presented in Fig. 3-6. Reactions under plasma conditions can cause formation of plasma species and final products that cannot otherwise be obtained by reactions under thermodynamic equilibrium. For example, species such as N_x with $x > 2$ [25], for nitrogen-containing plasmas, and Si_xCl_y with $x \geq 2$ [26], for plasmas containing silicon tetrachloride, have been observed only in cold plasmas. Special polymers have also been obtained by plasma polymerization (see Sec. 7.1).

Due to the multitude of the phenomena affecting the plasma and the resulting complexity of the plasma chemistry, it is not always possible to control the pathways of the chemical reactions or to predict from first principles the right combination of process parameters required for a certain outcome of the process. The kinetics of chemical reactions in a cold plasma can be considered as a

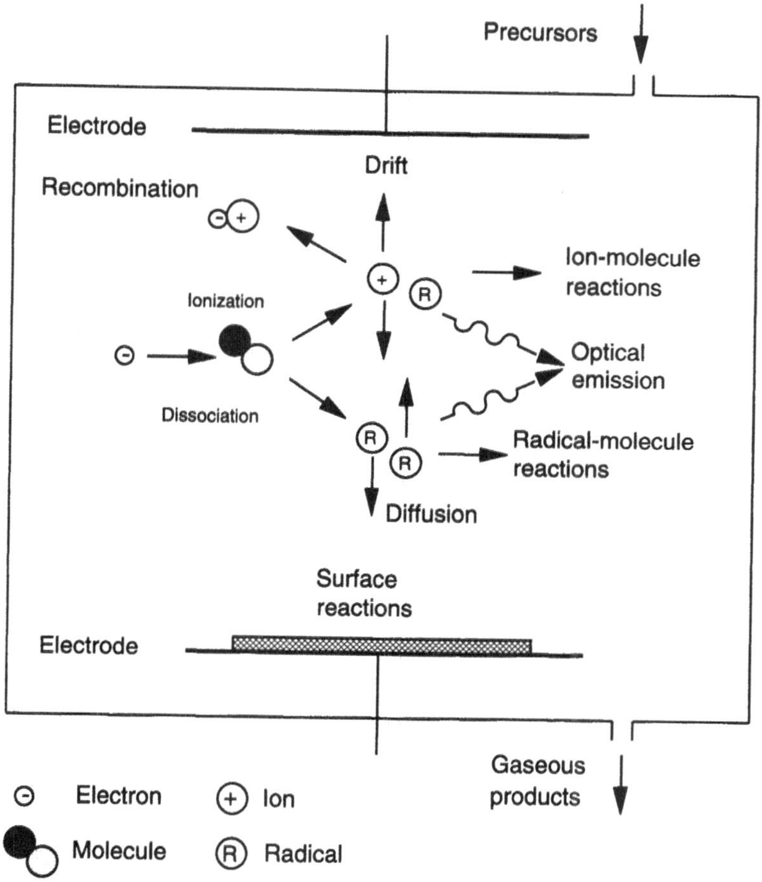

Fig. 3-6 Schematic diagram of reactions in a plasma reactor.

particular case of non-equilibrium chemical kinetics. While equilibrium kinetics can be explained by means of theories of elastic collisions, the kinetics of plasma chemistry also has to take into account inelastic collisions. In the absence of thermodynamic equilibrium in the cold plasma, the state of a plasma is determined by external parameters, such as electric power, pressure, gas flow rate, and internal parameters, that is, the rate constants of the reactions leading to formation or destruction of plasma species.

The knowledge of the kinetic constants is still insufficient for quantitative calculations of expected chemical behavior of most systems in the plasma. Yet semiempirical and semiquantitative models are used to predict the optimal conditions for given processes in the plasma. Plasma technology has reached the stage where it is possible to control empirically the right combination of parameters to obtain highly reproducible results, in the same system or between similar systems.

3.5 PLASMA SURFACE INTERACTIONS

The previous sections dealt with homogeneous chemical reactions occurring in the plasma bulk (Sec. 3.3.1) or heterogeneous reactions at the surfaces in contact with the plasma (Sec. 3.3.2). The existence of energetic particles and electromagnetic radiation in the plasma requires consideration of their interaction with the exposed surfaces. These physical interactions, which take place at any surface exposed to the plasma, have a significant effect on the results of the plasma treatment whether by affecting the heterogenous chemical reactions or the treated surfaces. The various particles of the plasma interacting with a solid surface exposed to it are shown schematically in Fig. 3-7.

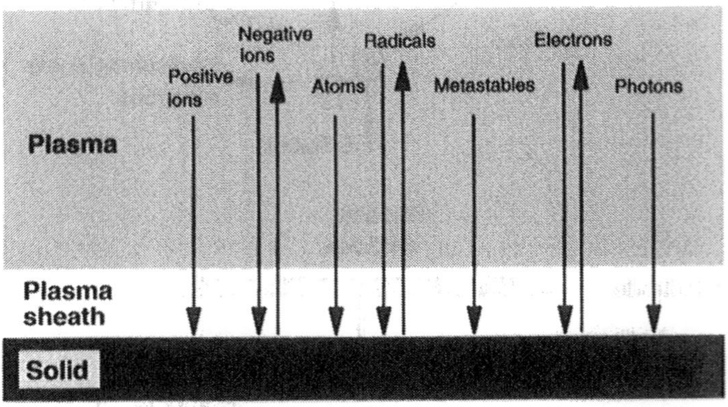

Fig. 3-7 Plasma particles interacting with a surface in a plasma.

The plasma contains energetic electrons and positive ions and also energetic neutrals. Most applications involve ions with energies in the range between 100 and 1000 eV. Bombardment of the solid surfaces with these particles can impart energy to the solid. When an energetic ion collides with a solid surface, the ion energy is transferred to the lattice atoms, creating a very-short-lived (10^{-11} to 10^{-12} sec) collision cascade of moving lattice atoms. The spatially very small, short-lived cascade can promote chemical reactions that either do not proceed at all or proceed very slowly under ambient conditions, at the temperature of the surface.

Upon contact with a surface, ions can release a significant part of their energy by producing pressure and thermal spikes. For example, thermal spikes of 3283 ^{0}K and pressures of 1.3×10^{10} Pa (1.2×10^5 atm) for 7×10^{-11} sec have been calculated for impinging ions with an energy of 100 eV [27]. According to Khaiat et al [28], it is possible to describe the neutralization of positive ions at a surface as short lived hot spots with a lifetime of the order of 10^{-14} sec. These energetic spikes obviously affect the results of the processes taking place at a surface in contact with a plasma and may be the reason for formation of *metastable materials*, such as amorphous hydrogenated silicon or diamondlike carbon. As a result of the bombardment, adsorbed species can be desorbed and substrate bonds may break causing crystal damage.

Different processes in the substrate or growing film that can be affected by the bombardment with the energetic particles are illustrated in Fig. 3-8 [29]. The figure presents the range of kinetic energies and flux densities existing in different

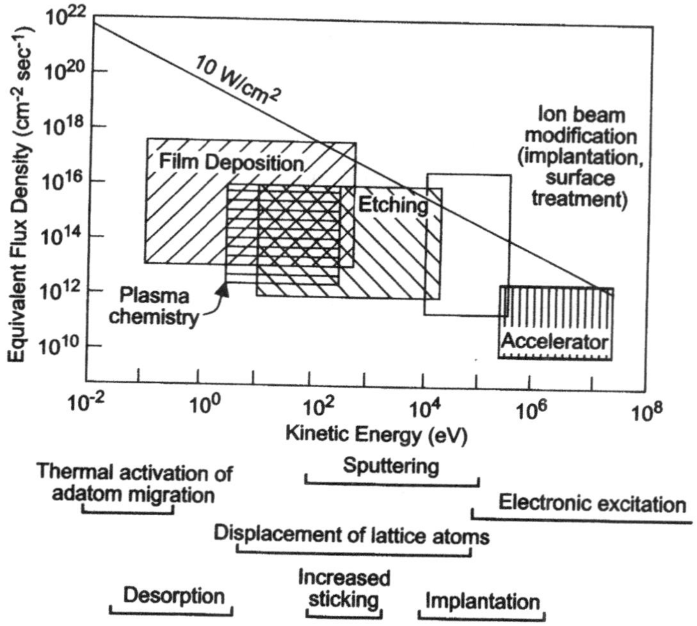

Fig. 3-8 Plasma surface interactions versus particle flux and kinetic energy (from [29], reprinted with permission).

plasma assisted techniques, while the processes caused by the bombarding particles in the surface layers of the solid are shown below the abscissa of the plot.

An early detailed description of the *plasma-surface* phenomena has been given by H.F. Winters [15]. According to Winters, the plasma-surface interactions can be divided into two categories. One type of interaction takes place relatively far from the surface in which the electronic excitation energy of the particles dominates. The second type of interaction requires a much closer proximity to the surface or even penetration of the particles from the plasma into the crystal lattice and, therefore, involves the kinetic energy of the particles. Neutralization of slow ions and associated electron emission fall into the first category, while cascade effects and sputtering fall into the second category.

Excited or ionized atoms can transfer their energy near a metallic surface by undergoing an electronic transition through three different mechanisms:

1. *Deexcitation* through emission of radiation

2. Deexcitation by a two-electron *Auger process*
3. A resonance process, whereby an electron is transferred from the metal to an equivalent level in the ion

A 100-eV He^+ ion spends about 10^{-13}–10^{-14} sec within 5 Å from the surface as compared to about 10^{-12} sec for a thermal particle. Deexcitation through emission requires much longer times, of the order of 10^{-8} sec. Therefore, the probability that the deexcitation occurs through emission of radiation is very small. Nevertheless, the probabilities of neutralization near a surface in contact with the plasma are very high, close to unity as a result of the Auger and resonance processes [15]. Theoretical investigations indicate that almost all incident ions will be neutralized within a few atomic radii of a substrate by electrons produced through Auger emission processes. As a result, the particles that finally strike substrates in the plasma are mostly neutral species.

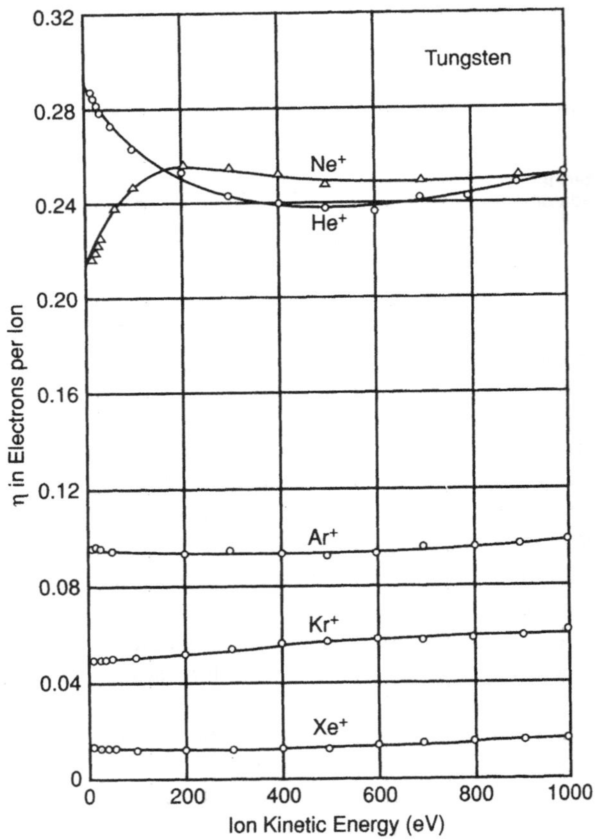

Fig. 3-9 Secondary electron yield from a tungsten surface versus ion kinetic energy (from [30], reprinted with permission).

The emission of Auger electrons leaves the solid substrate or growing film in an excited state while relaxation of the substrate from the excited state causes the generation of the secondary electron. As a result of the neutralization process, secondary electrons are emitted with a yield that is almost independent of the energy of the incident ion, but strongly dependent on its ionization potential. The *yield* is defined as the number of particles, in this case secondary electrons, emitted per arriving ion. As shown in Fig. 3-9, for a tungsten surface the yield of secondary electron emission is ~ 0.24 for He^+ and ~ 0.10 for Ar^+ ions arriving at the surface.

The emission of most of the secondary electrons is caused by the potential of the impinging ions (their electronic excitation energy), but if the kinetic energy increases above ≈ 1000 eV, *kinetic ejection* becomes dominant, and the yield for secondary electron emission increases with the kinetic energy of the ion. This is illustrated in Fig. 3-10, which presents the secondary electron yield for molybdenum, as a function of the kinetic energy of Ar^+ ions and Ar neutrals.

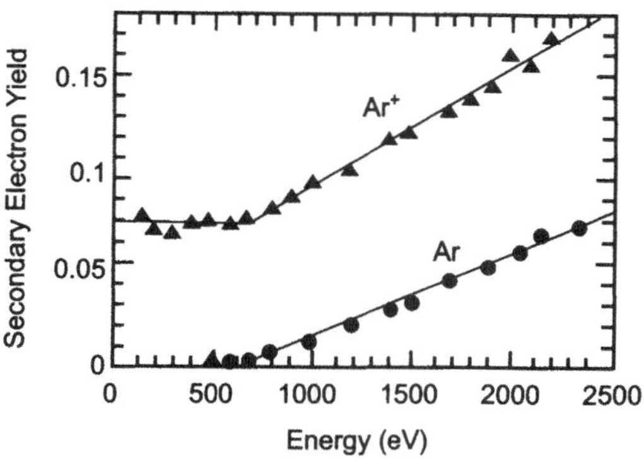

Fig. 3-10 Secondary electron yield as a function of kinetic energy for Ar and neutral atoms (from [15], reprinted with permission from S. Veprek and M. Venugopalan, eds., *Topics in Current Chemistry, Plasma Chemistry III*, Berlin: Springer-Verlag, 1980).

Ions and long-lived electronically excited metastable species of the plasma are deexcited with high probability at any surface in contact with the plasma causing concentration gradients of these species near the surface. The deexcited neutral particles are reflected from the surface with a significant fraction of their kinetic energy and with vibrational energy gained by multiatomic particles through the deexcitation process. Surfaces in the plasma are thus sources of energetic neutral particles and vibrational excited molecules.

After becoming neutralized, an ion contacting a surface exposed to the plasma begins to lose energy to the lattice of the solid. The energy can be lost in two ways:

- *Electron loss*, through electronic collisions by interactions with the electrons of the solid, resulting in excitation and ionization
- *Nuclear loss*, through the interaction of the incoming particle with the atoms of the lattice

For the ion energies encountered in a cold plasma, the electron loss is much smaller than the nuclear loss [15]. When an ion collides with a solid surface, it causes a cascade of moving atomic targets within the solid called a *collision cascade*. The nuclear loss occurs mainly through collision cascades which cause sputtering, *radiation enhanced diffusion*, backscattering of the incident ion, *ion induced chemical reactions*, or production of lattice damage.

The transfer of momentum from an energetic ion or neutral to the substrate causes heating, defect generation, and phonon interations in the surface layer. As a result, surface diffusion or adatom mobility is enhanced and improved step coverage can be obtained during film preparation by PECVD [31].

One effect of the nuclear loss is the *entrapment of gases* from the discharge in the deposited film. It was found that concentrations of argon entrapped in films can vary between 0.1 and 50 at% depending on the substrate bias. As illustrated in Fig. 3-11, significant amounts of argon were found in nickel films grown on

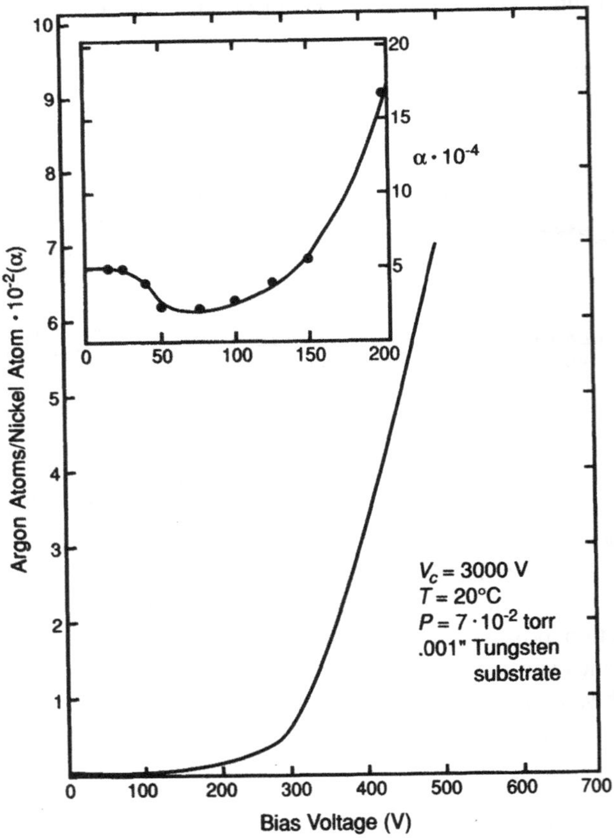

Fig. 3-11 The argon concentration in nickel films grown in a glow discharge versus bias voltage (from [32], reprinted with permission from *J. Appl. Phys.*, vol. 38, p. 3928, 1967).

substrates at negative biases bigger than 100 eV [32]. At such energies, the ions are accelerated toward the substrate and into the growing film where they are finally incorporated.

Sputtering is another important phenomenon occurring in a cold plasma. It is the ejection of material from a solid surface caused by the impingement of energetic ions. The sputtering yield, defined as the number of sputtered atoms ejected per each incident ion, depends among other parameters on the energy of the incident ion and on the ratio between the masses of the incident ion and sputtered atom. Sputtering is characterized by a threshold energy that depends on the impinging ion and the surface atoms. The sputtering threshold is in the range of 10–50 eV, as illustrated in Table 3-5 for elements bombarded with Ar^+ ions of different energies. Above the threshold energy, the sputtering yield increases with increasing ion energy.

TABLE 3-5 Threshold Values and Sputtering Yield for Argon Bombardment (from [33], reprinted with permission)

	Threshold (eV)	Ar^+ + energy (eV) 60	100	200	300	600
Ag	15	0.22	0.63	1.58	2.20	3.40
Al	13		0.11	0.35	0.65	1.24
Au	20		0.32	1.07	1.65	2.43
Be	15		0.074	0.18	0.29	0.80
Cr	22		0.30	0.67	0.87	1.30
Cu	17	0.10	0.48	1.10	1.59	2.30
Fe	20	0.064	0.20	0.53	0.76	1.26
Ge	25		0.22	0.50	0.74	
Mo	24	0.027	0.13	0.40	0.58	0.93
Nb	25	0.017	0.068	0.25	0.40	0.65
Ni	21	0.067	0.28	0.66	0.95	1.52
Pd	20		0.42	1.00	1.41	2.39
Pt	25	0.032	0.20	0.63	0.95	1.56
Re	35		0.10	0.37	0.56	0.91
Si			0.07	0.18	0.31	0.53
Ta	26	0.01	0.10	0.28	0.41	0.62
Ti	20		0.081	0.22	0.33	0.58
V	23	0.03	0.11	0.31	0.41	0.70
W	33	0.008	0.068	0.29	0.40	0.62
Zr	22	0.027	0.12	0.28	0.41	0.75

For conditions prevalent in most cold plasma processes, the negative self-bias of the surfaces in contact with plasma is sufficiently large to cause a certain amount of sputtering. The sputtered material can redeposit and contaminate the depositing film or the plasma treated surface.

Sputtering can also affect the plasma etching process, producing unwanted results. During plasma etching, nonetchable material sputtered from the walls of the reactor can deposit on the etched surface and prevent the etching of the area which is coated with the contaminants. As a result, surface roughening can occur or the etching reaction can stop completely [15]. The same effect can, however, be used to obtain wall tapering over steps during deposition of films by PECVD under plasma conditions that promote sputtering. Several aspects of plasma-surface interactions are discussed in greater detail in the following.

3.5.1 Ion and Electron Induced Chemical Reactions

In addition to the physical process described previously, the bombardment of a plasma treated surface with energetic particles can also affect the chemical reactions occurring at the surface and their rates. Reactions that take place at a surface exposed to the plasma can be affected by ultraviolet photons and soft X ray present in the plasma that are sufficiently energetic to break chemical bonds. However, electron and especially ion bombardments are the most effective in changing and promoting chemical reactions at the substrate. In addition to physical sputtering, ion enhanced chemistry is technologically probably the most important result of collisions of energetic ions with surfaces. Ion bombardment promotes the mixing of atoms near the surface by momentum transfer and by enhanced diffusion. This bombardment is partially responsible for the improvement in the quality of thin films that are deposited with simultaneous energetic ion bombardment.

It was also observed that surfaces subjected to ion bombardment etch much more rapidly than those kept near plasma potential. For example, in the XeF_2-Si-Ar system, gaseous XeF_2 reacts significantly with Si at room temperature [34]. But if the reaction takes place under simultaneous bombardment with ions, the probability that fluorine atoms will react to form the volatile SiF_4 and etch the Si is greatly enhanced.

This was proved by etching silicon separately with XeF_2, Ar^+ beam, and a mixture of $Ar^+ + XeF_2$. The results illustrated in Fig. 3-12 showed that the etch rate obtained with simultaneous use of XeF_2 and Ar was eight times higher than the sum of the etch rates obtained with each of the individual gases alone [35]. Similar enhancing effects produced by Ar^+ ions were observed for plasma etching of Si with F_2 or with Cl_2 and C etching with F_2.

Electron bombardment was also found to cause an enhancing etching effect. The bombardment of the surfaces exposed to the plasma by energetic electrons can also cause

- Emission of secondary electrons.
- Chemical reaction between gaseous species and the solid surface, when the reaction requires electron activation.
- Electron induced dissociation of adsorbed molecules: the bombardment with electrons of the species adsorbed on the surface can cause fragmen-

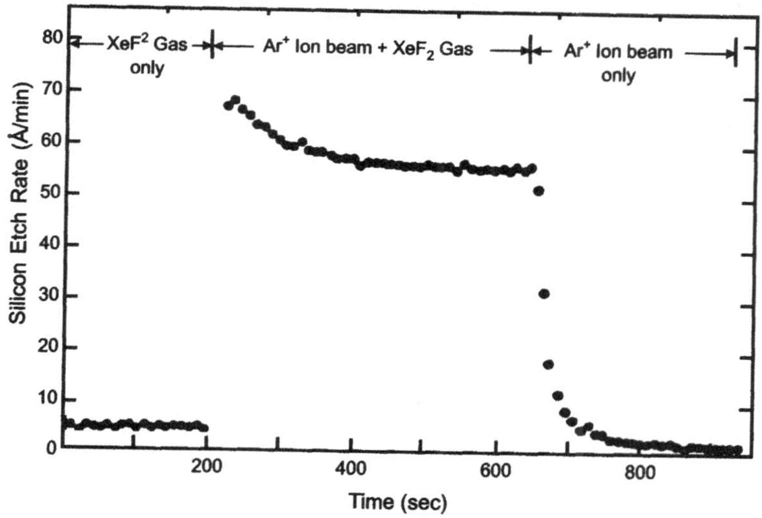

Fig. 3-12 Ion-assisted gas-surface chemistry using Ar^+ + XeF_2 on silicon (from [15], reprinted with permission from S. Veprek and M. Venugopalan, eds., *Topics in Current Chemistry, Plasma Chemistry III*, Berlin: Springer-Verlag, 1980).

tation of these species; some of the fragments will remain attached to the surface, while others will desorb into the gas phase; this process is called *electron stimulated desorption* (*ESD*).

- Lattice damage of sensitive materials.

Because the momentum of ions is much higher than that of electrons, the ion bombardment is much more effective in enhancing surface reactions than is electron bombardment.

To understand the enhancing effect produced by the impinging ions, one has to consider that steady-state etching of a solid exposed to a plasma or the film deposition go through several steps:

1. Generation of reactive species
2. Diffusion to the surface
3. Nondissociative adsorption of gas species to the solid surface
4. Surface reactions:
   a. Dissociation of adsorbed species (dissociative chemisorption)
   b. Reaction of adsorbed radicals with the solid to form adsorbed product molecules
5. Desorption of volatile species from the surface:
   a. Desorption of volatile products into the gas phase
   b. Removal of nonreactive residue from surface

6. Diffusion of volatile species away from the surface

The formation of a volatile product and its removal from the treated surface is essential to the etching process. Although in the deposition process the main product of the reaction is nonvolatile, secondary volatile reaction products may form at the surface. Removal of such volatile products and nonreacted species is important to prevent contamination of the growing film through their incorporation into it.

Steps 3 to 5 are heterogeneous stages that are affected by the substrate temperature, but their kinetics are not dependent solely on it. Bombardment of the plasma treated surface with energetic ions or electrons can affect each of these steps. For example, dissociative chemisorption may occur preferentially at defect sites produced by impinging ions. If such is the case, ion bombardment will strongly enhance the reaction rate. Ion or electron bombardment can also cause dissociation of adsorbed molecules, which would otherwise not take place.

If the formed product species are volatile, they will desorb, and the surface will be etched. If the reaction products are nonvolatile, a layer of reaction products will deposit on the surface. It may also happen that a monolayer adsorbed on the surface can prevent the dissociative chemisorption. For example, etching of Si by fluorine may be slowed by oxygen, carbon, or chlorine adsorbed on the surface. In such a situation, the cleaning of the surface through sputtering by energetic ion bombardment or through electron stimulated desorption will enable the dissociative chemisorption to take place and thus enhance the etching process. The enhancement of surface processes by ion or electron bombardment is a synergetic process.

The studies of the effects of surface bombardment by energetic particles pay special attention to etching, with emphasis on anisotropic etching. In anisotropic etching the removal of material from the plasma treated sample has to progress at a much faster rate in one direction than in the others. During processing of silicon wafers for VLSI, the dimensions of the individual lines become smaller, and anisotropic etching is required to prevent undercutting the structure underneath the mask. Anisotropic etching is also required to obtain deep trenches with straight walls.

Anisotropic etching can be obtained in most plasma systems by ion assisted gas-surface chemistry. As illustrated in Fig. 3-13, when the sample is maintained at a negative potential in plasma, positive ions bombard the surface at normal incidence, provided the dimensions of the feature are much smaller than the thickness of the plasma sheath, which is always the case in microstructure fabrication situations. Neutral species, in contrast, are incident randomly on the surface, and all surfaces are exposed to approximately equal fluxes of neutral species. Etching solely by neutrals will therefore cause isotropic etching and undercutting underneath the mask, as shown in Fig. 3-13 (a). As a result of the ion enhanced chemistry, the surface bombarded by the ions etches at a much higher rate than the lateral walls, which are not bombarded, and anisotropic etching with little undercut is obtained as illustrated in Fig. 3-13 (b).

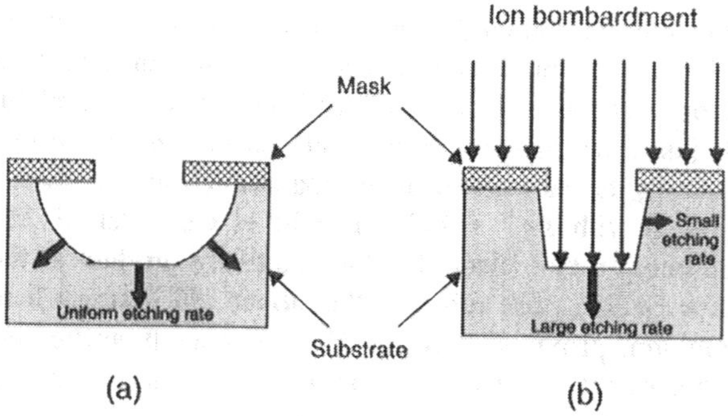

Fig. 3-13 Neutral versus ion assisted etching.

Let consider, for example, the *reactive ion etching* (RIE), of silicon by carbon tetrafluoride (CF_4). The plasma dissociates the inert fluoride into several reactive components:

$$e + CF_4 \rightarrow CF_3^+ + F + 2e \qquad (3.45)$$

The atomic fluorine that is believed to adsorb on the surface will etch the Si, though at a low rate. Bombardment of the surface with the CF_3^+ ions accelerates the etching rate. The acceleration of the etching by the CF_3^+ ions was explained in earlier proposals by assuming that the ions

- Provided additional F atoms to the etching process
- Accelerated the formation of SiF_2 by ion induced damage to the Si lattice
- Accelerated the desorption of SiF_2

However, the proposal that the accelerated reaction of F atoms with Si is caused by ion induced damage of the Si lattice is not supported by more recent experiments in which the etch rates of damaged and undamaged single-crystal Si were compared directly. Coburn et al. [21] separated the investigation of the adsorption from the following steps, that is, the formation of volatile product and the desorption of the volatile product. The authors first dosed the Si surface with fluorine, and then pumped away the fluorine from the gas phase before subjecting the fluorinated silicon surface to the ion bombardment.

Surface analysis by synchrotron photoemission of the fluorinated Si surfaces showed that the most abundant species on the surface was SiF_3 with very little SiF_4. On the other hand, SiF_4 molecules are found to be the predominant species evolved in this system during the ion assisted etching. The authors [21] therefore concluded that the ion bombardment promotes the formation of SiF_4 from the SiF_3 layer, which was already formed on the Si surface before the ion bombardment and its desorption from the surface. This conclusion was strengthened by the observation that when molecular fluorine was incident on the Si surface during bombardment with energetic ions, the etch rate observed was one order of magnitude lower compared to the etch rate observed with atomic fluorine flux. Molecular F_2 reacts with Si at much lower rates than atomic F.

The effect of the ion bombardment on plasma chemistry depends on the ratio between neutral and ion fluxes. For constant ion bombardment conditions, the etching anisotropy was found to decrease with increasing neutral flux.

The results just described emphasize that *neutral versus ion chemistry* has to be considered during the discussion of plasma-surface interactions. For example, during etching of Si with $Ar^+ + XeF_2$ a yield of the order of 25 Si atoms per Ar^+ ion was observed [21]. Since the dominant etch product in the Si-F system has been found to be SiF_4, this indicates that about 100 F atoms leave the surface for each arriving ion, provided that the XeF_2 supply is sufficiently large. The magnitude of this effect indicates that in this case the chemistry is provided essentially by the neutral species and not by the ions, because the amount of chemistry that can be provided by an ion is not sufficient to account for the observed etch rates and evolved SiF_4.

The relative importance of various ion induced phenomena can be very different for various chemical systems. For example, during etching of silicon with halide-containing gases, the diffusion of chlorine into silicon lattice is much less than that of fluorine. Consequently, ion induced mixing of the top few monolayers for enhanced reactivity is more important for the Cl-Si system than for the F-Si system.

There are also cases in which the rate of reaction between the incident gas-phase species is unaffected, or even retarded, by energetic ion bombardment, as for example the interaction of chlorine with aluminum or copper [22].

Synergetic phenomena similar to those described for etching can also be expected during deposition of films by PECVD. Creation of adsorption and nucleation sites, dissociation of adsorbed surface species, and promotion of chemical reactions due to particle bombardment can be expected especially with particle energies above 20 eV [29]. However, the different effects that the ion bombardment has on the primary processes occurring during film growth are extremely difficult to separate. The effects of the ion bombardment on the growing film as well as on the substrate are illustrated in Fig. 3-8.

3.5.2 Energy Transfer

Another aspect of the plasma-surface interactions is the transfer of energy. Energy transfer from plasma to solid surfaces occurs through optical radiation and fluxes of neutral particles and ions. The optical radiation has components in the infrared, visible ultraviolet, and sometimes soft X ray. When absorbed by a solid surface, the radiation usually transforms into heat. When the surface is a polymer, absorption of ultraviolet radiation can also break up the polymer to produce free radicals that can react with arriving plasma species.

The energy of the neutral particles is composed of kinetic, vibrational, dissociation (for free radicals), and excitation (for metastables) fractions. The dissipation of the kinetic and vibrational energy fractions causes heating of the substrate. The dissociation energy can also be dissipated through surface chemical reactions on polymers or trough surface reactions involving adsorbed species and recombinations on metals. The metastable species release their energy only through collisions that cause heating of metallic surfaces or the formation of free radicals on polymers [5].

The energy of the ion flux is composed of kinetic, vibrational, and electronic fractions. The ions are accelerated toward the surfaces exposed to the plasma that are negatively self- or externally biased relative to the plasma. The ion bombardment can affect the properties of the deposited film if the ion energy is sufficiently high. In an RF plasma, the kinetic energy with which an ion reaches the electrode depends on the ratio between the distance passed by the ion during half a cycle and the width of the plasma sheath. This distance is a function of the ion mobility, the sheath field, and the frequency of the electromagnetic field sustaining the plasma. If the distance traveled by the ion during half a cycle is larger than the width of the sheath, the ion can impact the electrode with energies equivalent to a DC discharge of a potential approximately equal to the RF amplitude. If the distance is much smaller than the sheath width, the equivalent DC voltage can be up to a factor of 3 less than that of the surface bias. On electrically floating surfaces, the ions arrive usually with energies up to 5 eV and can cause ablation or electrostatic charging of the surface. On biased surfaces the ions can arrive with kinetic energies of hundreds of volts and dissipate it by causing heating and sputtering of the surface.

The ionization energy is released by neutralization and is dissipated by either heating of metals surface or formation of free radicals on polymers.

3.5.3 Effects on Film Growth

The chemical reactions in the plasma and at the solid surface are determining the composition of the final product of a plasma process. However, in contrast to thermal systems in thermodynamic equilibrium, the plasma-surface interactions, in addition to enhancing the processes which take place at the surface, also affect the structure and properties of the plasma treated surfaces.

Ion impact that occurs on the negatively biased surfaces can affect the physical properties of the material. During deposition, it can cause densification and increased oxidation resistance of the films. It might minimize or eliminate columnar microstructures in metallization layers used in electronic devices. It can improve the quality of films grown under simultaneous ion bombardment, alter the state of stresses in the growing film, improve adhesion between film and substrate, or enhance conformal coverage [36]. Ion impact in PECVD can also cause enhanced diffusion, collisional mixing, and the formation of metastable materials, such as amorphous hydrogenated silicon or amorphous hydrogenated carbon (diamondlike carbon).

Following are a few examples that illustrate the effects of the plasma-surface interactions on the deposition of films and their properties.

- Enhancement of surface diffusion by ion bombardment during PECVD deposition of coatings on nonplanar surfaces results in improved step coverage [31, 37].
- The substrate temperature required to obtain crystalline coatings can be lowered using PECVD. Thus the amorphous to crystalline transition temperature could be lowered during deposition of silicon to ~ 300 °C [38] and single-crystal ZnO films have been obtained by PECVD at temperatures as low as 200 °C [39].
- Lowering the temperature of epitaxial deposition can be accomplished using low-energy ion bombardment during PECVD [40, 41].
- During deposition of diamondlike carbon by RF PECVD, polymerlike films are obtained on the grounded electrode, which is at low negative self-bias relative to the plasma, while hard diamondlike carbon films are obtained on the powered electrode, which is generally at larger negative bias.
- The deposition rate of diamondlike carbon on the powered electrode of a parallel plate reactor was found to be up to four times faster than that on the grounded electrode [42, 43].
- A similar effect has been observed in plasma polymerization where a higher rate of deposition was found on the powered electrode as compared to the one on the walls of the reactor [44, 45].
- It was also found that for both amorphous hydrogenated silicon and diamondlike carbon the anodic material has a higher hydrogen concentration and, consequently, a larger band gap than the cathodic material [46].
- Plasma nitriding of metals takes place at much higher rates and lower temperatures than does conventional thermal nitriding, and the composition and structure of the nitrided layers can be tailored by controlling the plasma process parameters.

3.5.4 Plasma Induced Damage

The bombardment of the substrate with energetic particles during cold plasma treatment can produce etch directionality and also required changes in the properties of the material. However, if the energies of the particles are high or the fluxes are large, damage can occur in existing films or substrates. Plasma induced damage can occur, especially during reactive ion etching when the substrate is negatively biased at relatively high bias of several hundred volts and is therefore exposed to bombardment of high-energy ions. The damage is caused through bond breaking by the bombarding particles and formation of defects in the surface layer of the substrate. Thus damage was observed in silicon wafers after reactive ion etching in CF_4 plasma, to which hydrogen was added to cause etch selectivity of oxide to silicon [47]. The damage was explained as being caused by physical bombardment of the wafer by hydrogen ions.

The study [47] indicated that the defect formation is not chemical but a result of bombardment of the surface with hydrogen ions. The damage consisted mainly of displaced Si atoms that migrated several nanometers and reorganized into interstitial dislocation loops. The defects are propagated during thermal oxidation into the substrate in the form of stacking faults below the oxide-silicon interface. The defects produced by the hydrogen-containing plasma caused the degradation of the lifetime of MOS capacitors. The degree of damage can be reduced by decreasing the exposure time to the hydrogen-containing plasma.

A similar study was performed on silicon wafers reactively ion etched in a $CF_4 + 40\%\ H_2$ plasma. The RIE was performed at a pressure of 25 mtorr and at a substrate self-bias of -425 V. The etching caused changes in the silicon wafer which consisted of [48]:

- A film containing carbon and fluorine, less than 50 Å thick deposited onto the silicon.
- A layer of ~ 30–50 Å heavily damaged, amorphized silicon was formed as a result of bombardment with heavy energetic particles which caused collision cascades.
- Hydrogen implantation (about 5%) to depth in excess of 200 Å; this layer was less damaged than the silicon layer immediately adjacent to the surface. Raman studies showed that the hydrogen is bonded to the silicon lattice.

A diagram of the damaged layer produced by the exposure of silicon to the plasma is shown in Fig. 3-14.

The changes caused in the material by ion bombardment resulted in the degradation of the electronic properties of the silicon as following:

- Degradation of the minority carrier lifetime
- Changes in the barrier height and ideality factor of Schottky barriers formed on dry etched silicon
- High contact resistance in contact hole etching
- Deterioration of the oxide quality (interface state density, dielectric strength) of thermal SiO_2 films grown on dry etched substrates

Similar behavior was observed in etching of silicon with fluorine containing gases. Fluorine was observed to a depth of 2000 Å with a maximum of 4% at a depth of 500 Å [49].

During ion bombardment, impurities from the plasma can penetrate the substrate either by implantation or through enhanced diffusion. Hydrogen atoms are especially prone to penetrate the substrate because of their small size. Ion bombardment can cause formation of point defects in the crystal lattice. The plasma can also cause contamination of the substrate with material sputtered from the walls of the reactor or the electrodes [50].

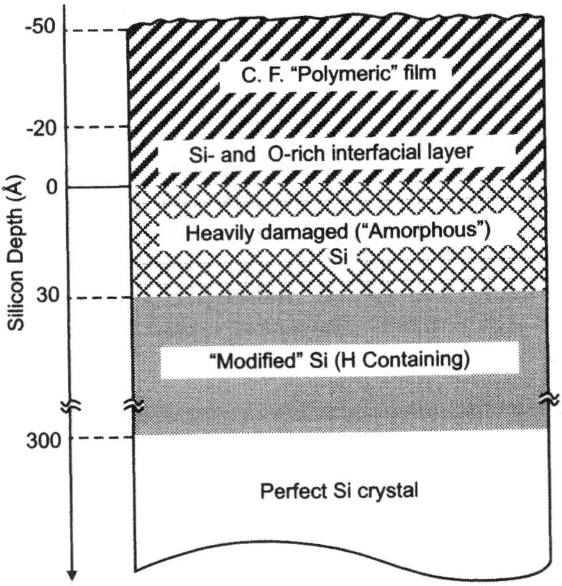

Fig. 3-14 Diagram of changes of Si near-surface region caused by $CF_4 + H_2$ reactive ion etching (from [48], reprinted with permission).

In addition to energetic ions and electrons the plasmas contain significant amounts of ultraviolet radiation, which can extend in energy to several tens of electron volts per photon. In dielectrics such as SiO_2 characterized by a bandgap smaller than 9 eV, the ultraviolet radiation can create defects that are detrimental to the surrounding circuitry. Positive fixed charge densities were observed in thermal oxides exposed to industrial plasmas of O_2 or SF_6, reaching values of 2×10^{17} cm^{-3} in a 1-kW microwave plasma at 2 mtorr [51].

Charge accumulation can occur on insulating layer, such as gate oxide, exposed to a cold plasma. At sufficiently high accumulation, the charges may cause electrostatic breakdown of the dielectric. Exposing a wafer containing MOS devices to a plasma discharge during patterning of polygate structures may cause damage to the device due to charge accumulation. It has been found [52] that at 13.56 MHz, the charge that can be transferred to the wafer is so small that it is unlikely that the gate will reach a level where electrical breakdown will occur. If, however, the frequency is low or a DC supply is used, breakdown of the oxides may appear. Samples subjected to a low-frequency or DC plasma were found to exhibit an increase in the density of states at the oxide semiconductor interface [52].

Recent studies [29] show that substrate damage is much smaller in films deposited by PECVD than in films deposited by sputtering. This may be due to the lower ion energies in PECVD plasmas as compared to those in sputtering plasmas. The difference in ion energies is caused by higher pressures and lower-power densities in PECVD. Substrate temperatures above 200 °C, generally used in PECVD, can also cause annealing of some of the defects during the deposition process. The substrate damage following plasma etching was also found to be greater than after plasma deposition. This may be a result of the fact that, while in PECVD the substrate is exposed to the plasma only for a short time at the start of the deposition, in etching the substrate is exposed to the plasma at the end of the process.

3.6 QUESTIONS

1. What type of reactions can cause dissociation of molecules in a low-pressure plasma reactor?
2. a. Describe the different ionization reactions that can occur in a plasma reactor.
   b. How are negative ions formed in a cold plasma and in what type of gases?
3. What reactions result in recombination of plasma species?
4. Which plasma reactions are associated with emission of electromagnetic radiations?
5. Describe the difference between ion-molecule and radical-molecule reactions. Which type usually dominates in cold plasmas and how can the overall process be affected by the relative contribution of each type of reaction?
6. Where do heterogeneous reactions take place in a plasma? Exemplify.
7. What processes take place at a surface in contact with a plasma? How can these processes cause contamination of a coating deposited in a plasma?
8. What type of material damage can be induced by exposure to a cold plasma?

3.7 REFERENCES

[1] Vursel, F., and L. Polak, In *Reactions under Plasma Conditions*, eds. M. Venugopalan, Vol. II, p. 299. New York: Wiley-Interscience, 1971.

[2] Beauchamp, J. L., and S. E. Buttrill, *J. Chem. Phys.*, 48: 1783 (1968).

[3] Venugopalan, M., ed., *Reactions Under Plasma Conditions*, Vol. I. New York: Wiley-Interscience, 1971.

[4] Venugopalan, M., ed., *Reactions Under Plasma Conditions*, Vol.II. New York: Wiley-Interscience, 1971.

[5] Bell, A. T., In *Techniques and Applications of Plasma Chemistry*, eds. John R. Hollahan and Alexis T. Bell, p. 1. New York: J. Wiley & Sons, 1974.

[6] Tal'rose, V. L., and G. V. Karachevtsev, In *Reactions under Plasma Conditions*, eds. M. Venugopalan, Vol. II, p. 35. New York: Wiley-Interscience, 1971.

[7] Vossen, J. L., and W. Kern, eds., *Thin Film Processes*. New York: Academic Press, 1978.

[8] McTaggart, F. K., *Plasma Chemistry in Electrical Discharges*. Amsterdam: Elsevier Publishing Company, 1967.

[9] Shahin, M. M., In *Reactions under Plasma Conditions*, ed. M. Venugopalan, Vol. II, p. 237. New York: Wiley-Interscience, 1971.

[10] Brown, S. C., *Basic Data of Plasma Physics*. Cambridge, MA: MIT Press, 1959.

[11] Jansen, F., In *Plasma Deposited Thin Films*, eds. J. Mort and F. Jansen. Boca Raton, FL: CRC Press, 1986.

[12] Manory, R. R., U. Carmi, R. Avni, and A. Grill, *Thin Solid Films*, 156: 79 (1988).

[13] Chatham, H., D. Hill, R. Robertson, and A. C. Gallagher, *J. Chem. Phys.*, 79: 1301 (1983).

[14] McDaniel, E. W., V. Cermak, A. Dalgarno, E. E. Ferguson, and L. Friedman, *Ion-Molecule Reactions*. New York: Wiley-Interscience, York, 1970.

[15] Winters, H. F., In *Topics in Current Chemistry, Plasma Chemistry III*, eds. S. Veprek and M. Venugopalan, p. 69. Berlin: Springer-Verlag, 1980.

[16] Fontijn, A., *Gas-Phase Chemiluminescence and Chemi-Ionization*. Amsterdam: Elsevier, 1985.

[17] Bell, A. T., In *Topics in Current Chemistry, Plasma Chemistry III*, eds. S. Veprek and M. Venugopalan, p. 43. Berlin: Springer-Verlag, 1980.

[18] Lotz, W., *Astrophys. J.*, Suppl. Nr. 128, XIV: 207 (1967).

[19] Compton, R. N., L. G. Christophorou, G. S. Hurt, and P. W. Reinhardt, *J. Chem. Phys.*, 45: 4634 (1966).

[20] Coburn, J. W., *IEEE Trans. Plasma Sci.*, 19: 1048 (1991).

[21] Coburn, J. W., *IBM Research Report RJ*, 7488 (1990).

[22] Winters, H. F., and J. W. Coburn, *J. Vac. Sci. Technol.*, B3: 1376 (1985).

[23] McFeely, F. R., J. F. Morar, and F. J. Himpsel, *Surf. Sci.*, 165: 277 (1986).

[24] Manory, R. R., R. Avni, and A. Grill, *Mat. Res. Soc. Symp. Proc.*, 54: 715 (1986).

[25] Polak, L., In *Reactions under Plasma Conditions*, eds. M. Venugopalan, Vol. II, p. 141. New York: Wiley-Interscience, 1971.

[26] Avni, R., U. Carmi, R. R. Manory, A. Grill, and E. Grossman, *J. Appl. Phys.*, 62: 2044 (1987).

[27] Tsai, H., and D. B. Bogy, *J. Vac. Sci. Technol.*, A5: 3287 (1987).

[28] Khaiat, Y. L., A. Inspector, and R. Avni, *Thin Solid Films*, 72: 249 (1980).

[29] Hess, D. W., and D. B. Graves, In *Microelectronics Processing*, eds. Dennis W. Hess and Klavs F. Jensen, p. 377. Washington, DC: American Chemical Society, 1989.

[30] Hagstrum, H. D., *Phys. Rev.*, 104: 317 (1956).

[31] Hess, D. W., *J. Vac. Sci. Technol.*, A8: 1677 (1990).

[32] Winters, H. F., and E. Kay, *J. Appl. Phys.*, 38: 3928 (1967).

[33] Venugopalan, M., and R. Avni, In *Thin Films from Free Atoms and Particles*, eds. Kenneth J. Klabunde, p. 59. London: Academic Press, 1985.

[34] Winters, H. F., and J. W. Coburn, *Appl. Phys. Lett.*, 34: 70 (1979).

[35] Winters, H. F., *J. Appl. Phys.*, 49: 5165 (1978).

[36] Denison, D. R., M. S. Chang, and T. Ebata, *Microelectron. Manufact. Test.*, 22 (July 1990).

[37] Rand, J. M., *J. Vac. Sci. Technol.*, 16: 420 (1979).

[38] Carlson, D. E., and R. W. Smith, *J. Electron. Mater.*, 11: 749 (1982).

[39] Shiosaki, T., T. Yamamoto, M. Yagi, and A. Kawabata, *Appl. Phys. Lett.*, 39: 399 (1981).

[40] Donahue, T. J., and R. Reif, *J. Appl. Phys.*, 57: 2757 (1985).

[41] Yew, T. R., and R. Reif, *J. Appl. Phys.*, 68: 4681 (1990).

[42] Vandertop, G. J., M. Kawasaki, R. M. Nix, I. G. Brown, M. Salmeron, and G. A. Samorjai, *Phys. Rev.*, B41: 3200 (1990).

[43] Catherine, Y., In *Diamond and Diamond-like Films and Coatings, NATO-ASI Series B: Physics*, eds. R. E. Clausing, L. L. Horton, J. C. Angus, and P. Koidl, Vol. 266, p. 193. New York: Plenum Publishing Corp., 1991.

[44] Smolinsky, G., and M. J. Vasili, *Int. J. Mass Spectrom. Ion Phys.*, 12: 47 (1973).

[45] Vasili, M. J., and G. Smolinsky, *Int. J. Mass Spectrom. Ion Phys.*, 18: 179 (1975).

[46] Mort, J., and F. Jansen, eds. *Plasma Deposited Thin Films*. Boca Raton, FL: CRC Press, 1986.

[47] Frieser, R. G., F. J. Montillo, N. B. Zingerman, W. K. Chu, and S. R. Mader, *J. Electrochem. Soc.*, 130: 2237 (1983).

[48] Oehrlein, G. S., R. M. Tromp, J. C. Tsang, Y. H. Lee, and E. J. Petrillo, *J. Electrochem. Soc.*, 132: 1441 (1985).

[49] Molchanova, S. A., Y. F. Nasedkin, A. A. Serov, and I. V. Kurchatov, In *Proc. 10th Symp. on Plasma Chemistry,* August 4-9, 1991, Bochum, Germany, eds. U. Ehlemann, H. G. Lergon, and K. Wiesemann, p. 2.2–15. Bochum: IUPAC, 1991.

[50] Oehrlein, G., In *Handbook of Plasma Processing Technology*, eds. Stephen M. Rossnagel, Jerome J. Cuomo, and William D. Westwood, p. 196. Park Ridge, NY: Noyes Publications, 1990.

[51] Devine, R. A. B., J. Pelletier, and Y. Arnal, In *Proc. 10th Symp. Plasma Chemistry*, Aug. 4-9, 1991. Bochum, Germany, eds. U. Ehlemann, H. G. Lergon, and K. Wiesemann, p. 2.2–41. Bochum: IUPAC, 1991.

[52] Ryden, K., and H. Norstrom, *J. Electrochem. Soc.*, 134: 3113 (1987).

4

Plasma Reactors

4.1 PLASMA SYSTEMS

A block diagram of a cold plasma system used for material processing is presented in Fig. 4-1. Generally, it consists of several subsystems providing different necessary functions as following:

1. Gas handling system, which includes

   a. *Precursor gas supply.* The source materials or precursors are in most cases gases in high-pressure cylinders or liquids with sufficiently high vapor pressures. Solids with reasonable vapor pressures are also used sometimes as precursors.

   b. *Mass flow controllers.* These are used to measure and control the flow of the different gases fed to the reactor.

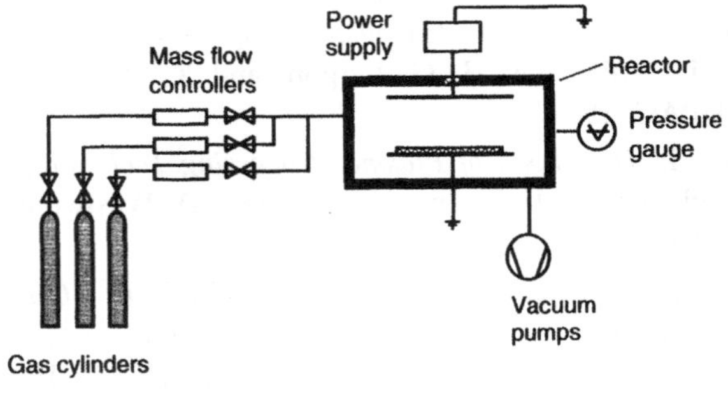

Fig. 4-1 General outline of a system for cold plasma processing.

c. *Vacuum system comprising pumps and pressure controller.* The plasma reactors for materials processing operate at pressures between 10^{-4} and ~ 10 torr. However, lower background pressure is often required to ensure the cleanliness of the process. Therefore, the whole range of vacuum pumps, from mechanical to cryogenic, are used; the types and sizes of the pumps are determined by the required vacuum levels and gas flow rates.

2. *Plasma reactor.* The plasma reactors will be described in detail in later sections of this chapter.
3. *Powers supplies.* Different types of power supplies will be discussed together with the corresponding reactors. The role of power supplies is twofold: to sustain the plasma in the reactor and to provide a controlled external substrate bias when such is needed for the process.
4. *Safety devices* for handling hazardous gases. Most of the precursors used in plasma processing are hazardous, some even extremely so; these gases can by corrosive, highly toxic, flammable, or explosive. Some gases, such as the extensively used silane, are unpredictably pyrophoric. Germane, or doping gases, such as diborane, phosphine, or arsine, are extremely toxic. Some of the safety devices required when dealing with such hazardous materials are

a. *Flow limiters*, mounted between the valve of the supply cylinder and the pressure regulator to prevent excessive flow of the gas in case of breakdown of the pressure regulator.

b. *Flashback arresters*, required when using flammable or explosive gases to prevent fire propagation to the gas in the cylinder; located in front of the pressure regulator.

c. *Cross-purge assemblies*, used to purge the regulators and prevent release of the hazardous gas to atmosphere when exchanging the gas cylinders; they are mounted between the cylinder and the pressure regulator.

d. *Scrubbers or diluters* of exhaust from the pumps.

e. *Detectors* for hazardous gases.

4.1.1 Process Parameters

For a given system, the outcome of the process is strongly dependent on its parameters. A comprehensive list of the process parameters that can affect the end result of plasma processing is given in Table 4-1 [1]. The classification of these parameters largely complies with the first three subsystems described before. The following list names the most important parameters among those in Table 4-1:

- Partial pressures of the feed gases, or flow rates of the different gases
- Total pressure in reactor
- Substrate temperature and bias

TABLE 4-1 Parameters Controlling Materials Processing by Cold Plasmas (adapted after [2], reprinted with permission)

Plasma Processing Parameters		
Kinetic (gas system)	**Electrical (plasma system)**	**Surface (substrate system)**
Precursor gases	Frequency (DC to GHz)	Material
Carrier gases	Free fall	Conducting
Mass flow rates	Mobility	Insulating
Pressure	Diffusion	Temperature
Gas delivery location	Electrode geometry	Position
	Electrodeless	
	Electrode	
	Discharge power	
	Field strength	
	Current density	
	Particle energy	
	Active neutrals	
	VIS to UV radiation	
	Electrode material	

- Reactor geometry and material
- Electrode material and distance between electrodes
- Electric power applied to the plasma

These parameters have to be carefully controlled to define the plasma chemistry and to achieve the desired results.

The complexity of the relationship between the macroscopic variables of a cold plasma processing system and the microscopic plasma parameters is illustrated in Fig. 4-2. Changes in macroscopic plasma variables such as composition of feed gas, flow rates, pumping speed, electrical discharge power, and frequency will generally change the basic plasma conditions, but the precise manner of these changes is in most cases unknown.

The RF power determines the current and voltage between the electrodes. The composition of the gas mixture can affect the chemical reactions inside the plasma reactor and the properties of the final product, for example the composition of deposited film or the anisotropy and selectivity of etching. The composition of the gas feed, the total pressure in the reactor, and the electrical power will each affect the rates of the reactions. The reaction rates can be affected also by the distance between electrodes and by substrate bias or temperature. These last two plasma parameters can also affect the structure of a deposited film or the etching

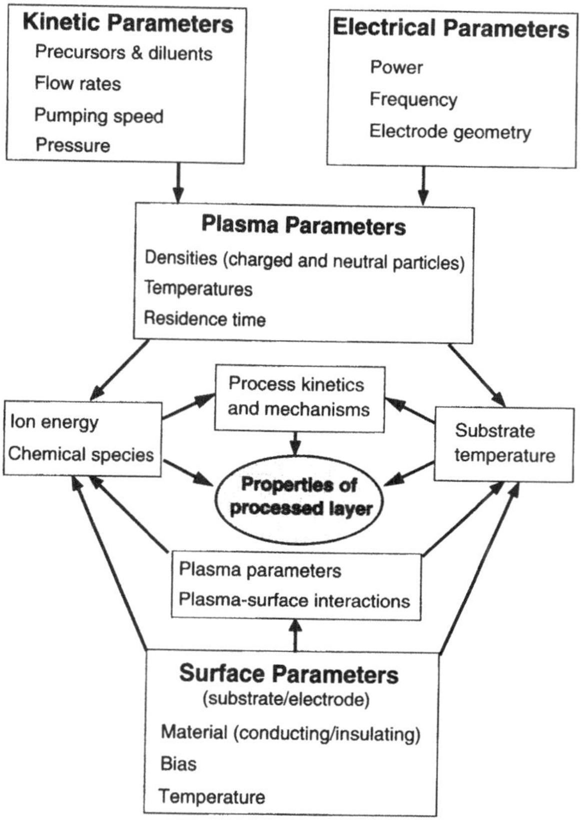

Fig. 4-2 Complexity of interaction between plasma variables (after [3], reprinted with permission). The arrows indicate the possible interactions between process parameters; they do not indicate that all the described interactions always take place.

anisotropy. The flow rates, total pressure, and reactor geometry can determine the uniformity of the process.

The gas flow rate, the pumping speed, and the pressure are interrelated. The pressure can be changed in two ways: by changing the flow rate at constant pumping speed or by changing the pumping speed at constant flow rate. While the two methods can provide the same pressure, they result in different residence times for the species in the reactor (see Sec. 4.1.2). The different residence times can cause changes in the chemistry of the process [4].

The frequency of the electric field is also a system parameter that can affect the plasma process or the properties of the product. A variation of the frequency can affect the number and energy of ions that can follow the alternating field, thus changing the flux and energy of the particles bombarding the plasma treated surface.

The process parameters must therefore be precisely controlled to achieve the required results and process uniformity.

And last, but not least, the building materials of the reactors and electrodes are an important factor to be considered for a specific process. Electrode and reactor materials can change the chemistry occurring in glow discharge. The materials should be chemically inert to the precursor gases and to the plasma products. Because surfaces in contact with a plasma are always negative relative to it, the electrodes and reactor walls will almost always be sputtered to a certain degree. The sputtered atoms may contaminate the plasma and the processed sample. The sputtered atoms may be incorporated in a deposited film or may mask areas of an etched surface causing nonuniform etching. Such possible effects should be taken into account when designing a reactor.

4.1.2 Scale-up

The transfer of a plasma process from a research laboratory apparatus to commercial manufacturing is called *scale-up*. The scale-up requires the upgrading of the working laboratory process to a technology that can process uniformly larger samples in commercial quantities. The scale-up should take into account economical considerations besides overcoming technological difficulties, some of which will be discussed in the following paragraphs. Because commercial scaled-up reactors are very expensive, it is of utmost importance to make a successful transfer from a small-scale process into a large one.

As a plasma process can operate in a large parameter-space and due to the complexity of interaction between the parameters (see Fig. 4-2), it is important to understand, at least to first approximation, how to adjust the plasma parameters according to changes in the size of the reactor.

The substrate temperature will normally be kept unchanged during the transfer of the process from a small to a large reactor. The main question is how to adjust gas flow rates, pressure, and electrical power in such a way that the scale-up of the reactor will not significantly affect the plasma chemistry. For this purpose, plasma processing can be conceptualized as being constituted of two subsequent processes [5]:

1. Formation of reactive species
2. Mass transport of the reactive species to the processed surface by convective diffusion

For the first process, it has been shown theoretically that, if the dissociation of the precursor molecules takes place through single collisions, the rate coefficients for the reactions caused by electrons depend only on the value of the ratio E/p, because the electron temperature in the plasma is determined mainly by E/p (E is electric field strength). One may therefore consider as a first approximation the fact that molecular dissociation rates caused by electron impact are dependent only on the E/p ratio. However, if dissociation is produced through multiple collisions, the dependency on E and p is more complex. It is therefore usually not sufficient to keep only the E/p ratio constant when scaling-up the process.

Because the product $D \cdot p$ (D being the diffusion coefficient) is constant for a gas, the mass transport through diffusion is also pressure dependent. Therefore, to preserve both the plasma chemistry and the mass transport during scale-up, one has to keep both p and E unchanged during scale-up. To maintain the same electrical field strength across the discharge over the larger area of the manufacturing reactor, it is required to increase the total current or power supplied to the discharge. In a parallel plate reactor, where the plasma is usually confined between the electrodes, the power has to be scaled linearly with the electrode area to keep the electric field constant. In other types of reactors, especially in tubular ones, the plasma volume does not coincide with the volume between electrodes and is affected by both pressure and flow rate. The power of the scaled-up reactor has to be therefore readjusted, often experimentally.

With the electric field, pressure, and substrate temperature kept constant during scale-up, the only external adjustable parameter left is the gas flow rate. The average gas flow velocity determines the convective mass transport of the reactive species to the substrate. The flow rate should be scaled up in such a way as to keep the average flow velocity identical in small and large reactors. It can be shown that this requirement regarding the flow velocity scales the gas flow rates proportional to the cross-sectional areas of the reaction zones, perpendicular to the gas flow direction [5].

A different approach is to keep the same average gas *residency time*, τ_r, in both large and small reactors. The residency time is defined as

$$\tau_r = \frac{p\, Vol_r}{Q} \tag{4.1}$$

where Vol_r = volume of reaction zone
Q = total mass flow rate

The gas residency time is a measure of the average distance over which the reactive species diffuse in the reaction zone, and thus determines the diffusional mass transport. According to Eq. (4.1), if the pressure is kept constant, the requirement of constant residence time scales the gas flow rates proportional to the volumes of the reaction zone. If the interelectrode distances are the same in the large and small parallel plate reactors, the requirements for the same gas velocity and same gas residency time are equivalent. In other types of reactors, the volume of the reaction zone can be dependent on the gas flow rate, and this approach cannot be used.

In summary, the main process parameters have to be adjusted during scale-up to a first approximation as following:

sample temperature	unchanged
pressure	unchanged
electric power	proportional to area of electrodes in parallel plate reactors

gas flow rates	proportional to electrode area normal to the flow direction, or proportional to volume of reaction zone.

It thus appears that the main changes in plasma parameters during scale-up are the linear extrapolations of electric power and gas flow rates. However, these changes alone may not be sufficient. In practice, a change in reactor geometry may affect other important intrinsic plasma parameters, especially if the reactors are not of the parallel plate type.

The foregoing description of scale-up principles did not take into consideration how the active species are formed along the flow path of the gas from the gas inlet to the treated substrate and their interaction before reaching the substrate. The formation and transport of the reactive species that form the final product are dependent on the flow patterns and on the time spent along the path to the substrate. The diffusion of the precursor molecules takes place along most of the distance from the gas inlet to the substrate, while diffusion of reactive species takes place mainly from the boundary of the plasma zone to the substrate. The reactor geometry and the geometry of the electrodes (size, distance between them) can change the gas flow patterns, the relative volume of the plasma region, the transport of the reactive species, and the rate and uniformity of the process.

Electric field patterns, especially at the edges, are also influenced by the reactor geometry and may affect the uniformity of the process at the edges of the substrate. This is especially true for tubular reactors or for bell jar reactors, often used in plasma polymerization [6]. In such reactors the volume of the plasma is not defined by the electrodes and is dependent on the interaction between different plasma parameters. Finally, the wall temperature of the scaled-up reactor is generally different from that in the smaller one, causing different gas retention or memory effects.

Because of these effects, the approximated parameters of the scaled-up reactor have to be generally readjusted either empirically or using theoretical modelling of the process and reactor. The last approach would generally still need empirical fine-tuning of the process parameters.

The ease with which a reactor can be cleaned has to be given careful consideration during its design. This is especially important for deposition reactors. Film material that deposits and accumulates on the surroundings of the deposition zone and on the walls of the reactor has to be regularly removed from any deposition system to avoid flaking and powder formation that may have adverse effects on the quality of the film deposited on the substrates.

Another important aspect to be considered when scaling up a laboratory system to a commercial manufacturing unit is the efficiency of precursor utilization, or the *material efficiency* of the process. In a deposition process, the material efficiency can be defined as the ratio between the weight of the deposited film and the weight of the film components in the feed gas. The material efficiency can be defined similarly for an etching process, in this case taking into account the removed instead of the deposited material. The highest material efficiency is generally obtained at high plasma density and low flow rates and, at a first approximation, is independent of pressure for the processes described in this book (based on gaseous precursors). The pressure, however, affects the process efficiency in sputtering processes, where the sputtering yield is pressure dependent. Because of the condition of low flow rate, high material efficiency is usually associated with lower process rates in deposition and etching. As a result, high material efficiency does not always imply maximum process efficiency, which demands high process rates [5].

The optimum trade-off between reactor design, material efficiency, and process rate is determined by each individual manufacturing process. The development of a manufacturing process is finally to a large extent empirical in defining the optimal conditions for achieving either film deposition or etching rate uniformity.

Due to the diversity of plasma reactor designs and plasma parameters, it is not within the scope of this chapter to describe in detail all designs and experimental approaches used. Only the general principles of the most common used reactor designs for producing the plasmas, based on the methods discussed in Chapter 2, are described in the following.

4.2 DC REACTORS

DC plasma reactors operate as described in Sec. 2.1. The plasma is sustained between two parallel plate electrodes as illustrated in Fig. 2-1, and the electric power is supplied to excite the plasma in the abnormal discharge mode. The distance between the electrodes and the pressure in the reactor have to satisfy Paschen's law for the used gas mixture (see Sec. 2.1.1).

The power supplies used to sustain DC plasmas can generally be used at constant voltage, constant current, or constant power mode. The power supplies should be able to control the preset value in each mode. The excitation and sustainment of a DC plasma require the use of electrically conductive electrodes and samples. However, electrically insulating films may deposit locally on the surface of the electrodes or samples due to contaminants formed in the plasma. In such cases, local dielectric breakdown normally occurs, causing arcing that results in spikes of high currents.

The power supply must be able to withstand these spikes and return to normal operation without shutting off. However, it must also be able to distinguish between the transient arcing currents and shorts that may occur between the electrodes and grounded surfaces. This feature, called *arc suppression*, must be a characteristic of a DC power supply of a plasma reactor.

4.3 RF REACTORS

The power supplies used to sustain commercial RF plasma systems are operated at specific frequencies established by international agreements [7]. The most used frequency is 13.56 MHz, but sometimes the higher harmonics of 27.12 MHz and 40.68 MHz are also used. The RF generators are designed to operate at constant output impedance of 50 Ω. The impedance of the glow discharge is, however, much higher and varies with the process parameters. An impedance matching network or matching unit is thus required between the RF power supply and the reactor. Otherwise, most of the power will reflect to the power supply instead of being absorbed into the plasma.

The matching unit is placed physically as close as possible to the powered electrode. In parallel plate reactors the matching unit is usually mounted on the flange supporting the powered electrode. Normally, variable automatic matching units are used, and these adjust themselves to keep the output impedance of the RF power supply constant at 50 Ω, independent from the changing plasma conditions.

4.3.1 Electrodeless Discharges

When using high-frequency power supplies, it is possible to transfer the energy to the plasma without inserting the electrodes inside the reactor, if a suitable coupling between the power supply and the reactor is used. For such a design, the reactor has to be built from dielectric materials such as quartz or Pyrex, which are generally used for this purpose. Since the electrodes are not in direct contact with the discharge, the plasma formed is called an *electrodeless discharge*.

Electrode designs for transferring the power to the plasma in electrodeless discharges are illustrated in Fig. 4-3. These electrodes correspond to two types of coupling the RF power to the electrodeless plasmas:

1. *Inductive coupling*, in which a coil connected to the power supply through a matching unit is wound around a tubular reactor, as shown in Fig. 4-3 (a).
2. *Capacitive coupling*, in which the RF power is transferred between two separated electrodes mounted outside the reactor. The electrodes can be in the shape of two rings mounted around the reactor as shown in Fig. 4-3 (b) or two plates placed along the reactor as shown in Fig. 4-3 (c).

Although the terms of inductive and capacitive coupling are often used as just defined, the inductive coupling is not purely inductive, and it always has a capacitive component too, through the wall of the reactor. Thus, when an inductive coupling is used, deposition on the walls is often observed to follow a pattern matching the shape of the coil. This is an indication of localized stronger electric fields on the walls, showing that the coupling is at least partly capacitive through the walls of the reactor [2].

In a tubular reactor of one of the types shown in Fig. 4-3, and especially in Fig. 4-3(a), the strength of the electric field is not uniform along the reactor. Therefore the substrate position inside the reactor becomes a parameter that can

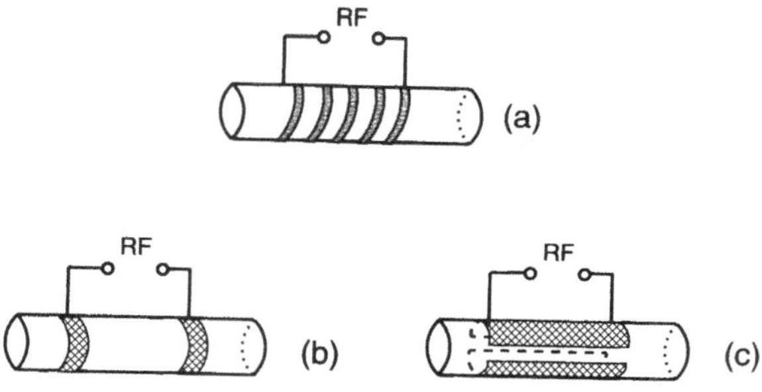

Fig. 4-3 RF coupling methods to electrodeless reactors.

affect the result of the process and can be controlled to achieve desirable properties [8, 9]. In tubular reactors the substrates are usually positioned with their surfaces perpendicular to the axis of the reactor.

The types of RF reactors just described, also called *barrel reactors*, are used sometimes for industrial applications, although they are more popular in research laboratories, being relatively inexpensive and easy to set up. In industrial applications, the tubular reactors are normally used for ashing of photoresist or etching in microelectronic fabrications and for noncritical deposition processes where uniformity is of no great concern.

A barrel reactor used for etching often includes an *etch tunnel*. The metallic perforated cylindrical etch tunnel, illustrated in Fig. 4-4, is inserted in the reactor

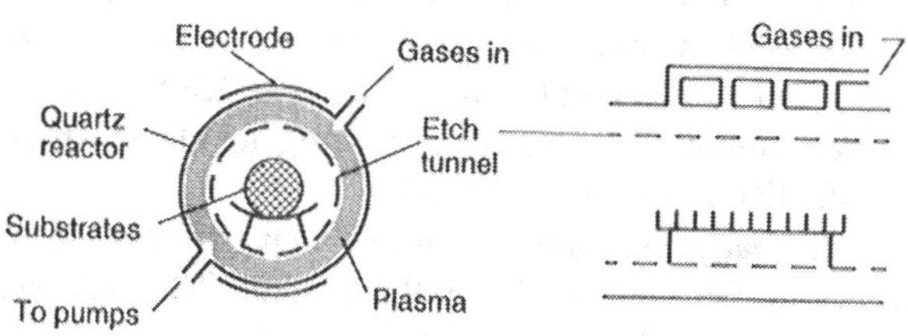

Fig. 4-4 Diagram of a barrel reactor.

to improve temperature uniformity along the length of the sample holder and to minimize particle bombardment of the substrates. This metallic cylinder acts as a Faraday cage and confines the glow region to the annulus between the etch tunnel and the wall of the reactor. Substrates are thereby shielded from direct contact with the plasma and are subject to little, if any, ion or electron bombardment. Neutral species diffuse, nevertheless, through the perforations and reach the surface of the processed substrates.

4.3.2 RF Discharges with Electrodes

4.3.2.1 Diode Reactors

The most common approach in the design of industrial RF plasma reactors is to couple the RF power to two parallel electrodes inserted inside the reactor. The coupling is done through special RF vacuum feedthroughs. One electrode is often grounded together with the walls of the reactor, as shown schematically in Fig. 4-5. This arrangement is also called the *diode* or *parallel plate reactor*.

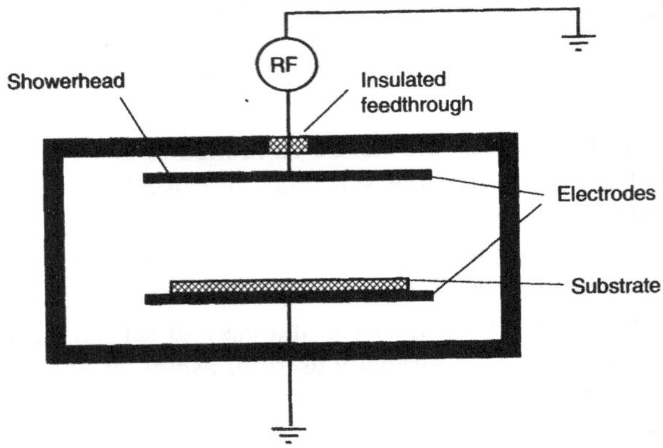

Fig. 4-5 Power coupling to a planar diode reactor.

For RF discharges, the surface of the electrodes in contact with the plasma are no longer required to be electrically conductive since RF voltages can be coupled through any kind of impedance. The surface of the electrode, substrate, or deposited coating can be electrically insulating. The RF power is, nevertheless, supplied in all cases through a conductive backing to the electrodes, to ensure uniform electric potential across the electrode.

Although the discharge is mainly confined between the electrodes, plasma species also diffuse to the side walls of the reactor. The interaction between plasma and electrodes can result in contamination of the plasma with atoms, which are either physically sputtered or chemically etched from the electrodes. The same is true, though to a lesser extent, for contamination with atoms from the walls. The right choice of materials for electrodes and reactor is therefore of utmost importance, especially in microelectronic processing, for which prevention of contamination is imperative. The electrodes are usually made of stainless steel and the reactors are built from stainless steel, aluminum, or quartz.

Capacitive coupling to parallel electrodes inside the reactor enables the creation of uniform electric fields, making it possible to achieve high process uniformity over large areas, determined by the size of the electrodes. However, the uniformity of the electric field alone is not sufficient to guarantee the uniformity of the results of PECVD processing when reactive gases are involved. Depending on their concentration and consumption and on their flow pattern in the reactor, the gases can be depleted of the reactive components along their path. This depletion can cause nonuniform deposition or nonuniform etching across the surface of the substrate.

The *radial flow reactor*, originally developed by Reinberg [10, 11], was designed to allow compensation for the depletion of the gases from the reactive components, thus improving process uniformity. The precursor gases were introduced at the periphery of the sample electrode and then flowed inward where they were exhausted at the center as shown schematically in Fig. 4-6. The discharge intensity is highest in the central region, promoting faster depositions in that area.

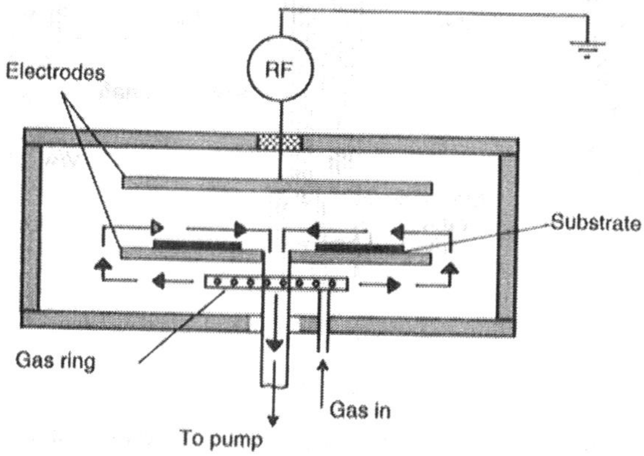

Fig. 4-6 Diagram of a Reinberg reactor.

The higher velocity and consequent shorter residence time of the gases at the center of the electrodes can counterbalance the higher discharge intensity in the center, leading in many cases to a uniform deposition rate across the whole radius.

A reactor similar to the Reinberg reactor used for PECVD processing is shown in Fig. 4-7 [12]. In this case, however, the gas enters through the center of the lower electrode and flows out at the periphery. Another often-used approach for achieving uniformity over a large area is by feeding the gas through a showerhead that forms one of the electrodes in the parallel plate reactors. The sample electrode can sometimes be heated or rotated to smooth out nonuniformity in processing.

4.3.2.2 Triode Reactors

If both electrodes in a parallel plate RF reactor are insulated from the walls, the reactor is called a *triode*-type *reactor*. In the triode reactor, each electrode and the reactor walls can be powered independently, and each can be biased, floating, or grounded. A diagram of a triode reactor is shown in Fig. 4-8.

Another version of the triode reactor is the one in which the discharge is sustained between two electrodes and the substrate is supported on a third elec-

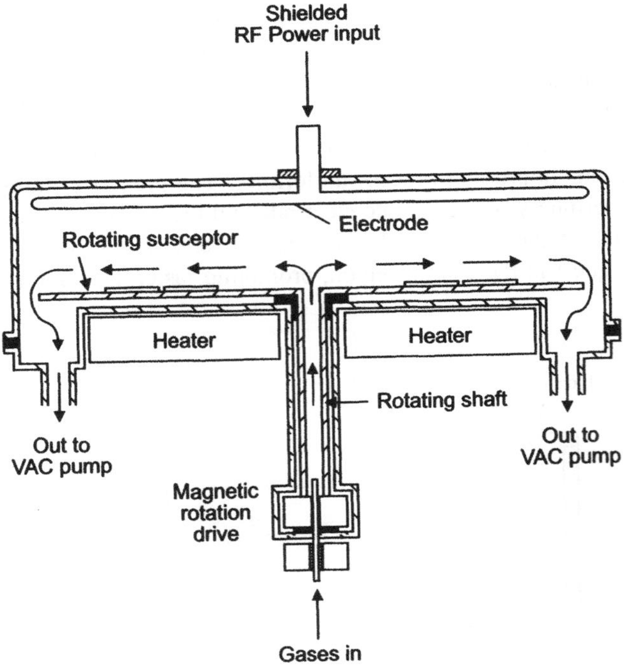

Fig. 4-7 Diagram of a radial flow reactor (from [12], reprinted with permission from *Thin Solid Films*, vol. 113, p. 135, 1984).

trode, which can be biased separately [13]. The electrodes in a triode reactor can be powered by combinations of power supplies, such as RF and microwave, RF and DC, RF and AC (< 50 kHz). While in the diode reactors, the RF power density may be limited by the maximal bias allowable on the substrate electrode, this limitation is eliminated in the triode reactors. The plasma density is controlled

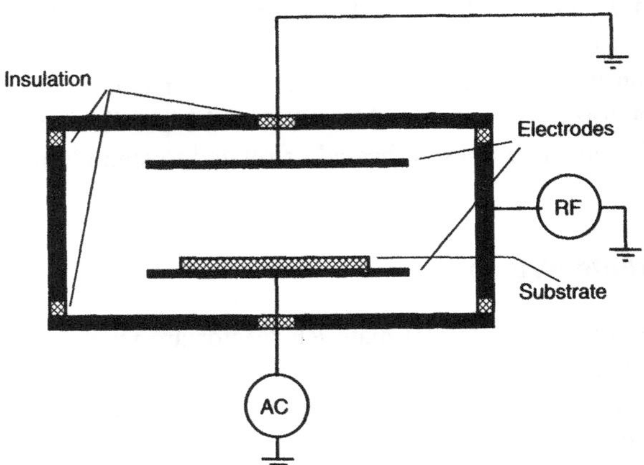

Fig. 4-8 Power coupling to a planar triode reactor.

by the power sustaining the plasma, while the substrate bias and thereby the energy of the ions bombarding it is controlled directly by the generator supplying the power to the substrate electrode.

Sometimes the discharge in a planar reactor may be assisted by thermionic electrons emitted from a hot filament; this type of reactor is also called a triode reactor.

4.3.3 High-Density RF Reactors

While different variations of the RF reactors described previously are widely used in manufacturing, efforts are made to develop *high-density RF* plasma *reactors* for achieving enhanced process rates. Two such high-density RF plasma sources are the *helical resonator* and *helicon* or *"whistler" source* [14]. In the helical resonator reactor shown in Fig. 4-9, the plasma is excited by a slow wave structure that achieves quarter- or half-wave resonance in the source [14]. The helicon source uses magnetic fields to obtain whistler mode waves that are launched into the reactor chamber. The helicon reactor is illustrated in Fig. 4-10.

The helicon plasma reactor is composed of a glass plasma source, to which electrical power is coupled inductively, and a metallic reaction chamber. The helicon source comprises an external RF antenna consisting of two loops diametrically placed on the outside of the glass tube and an electromagnet that produces a constant axial field. This combination of the RF field with the magnetic field excites a helicon wave in the source tube. Energy transferred from the helicon wave sustains the plasma. The axial magnetic field restricts the radial motion of the electrons and confines their movement along the magnetic field lines. If the

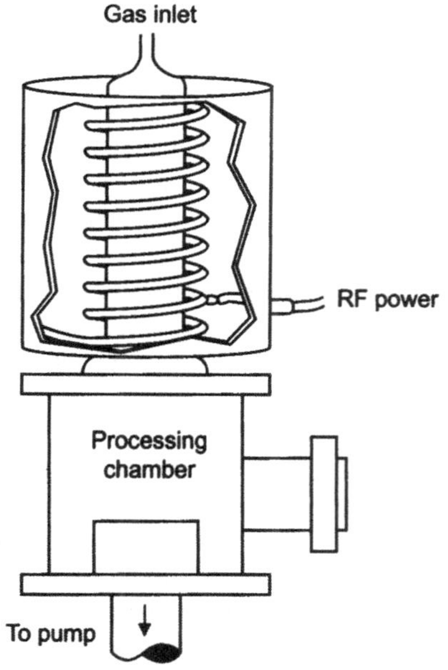

Fig. 4-9 Diagram of plasma reactor with helical source (from [14], reprinted with permission).

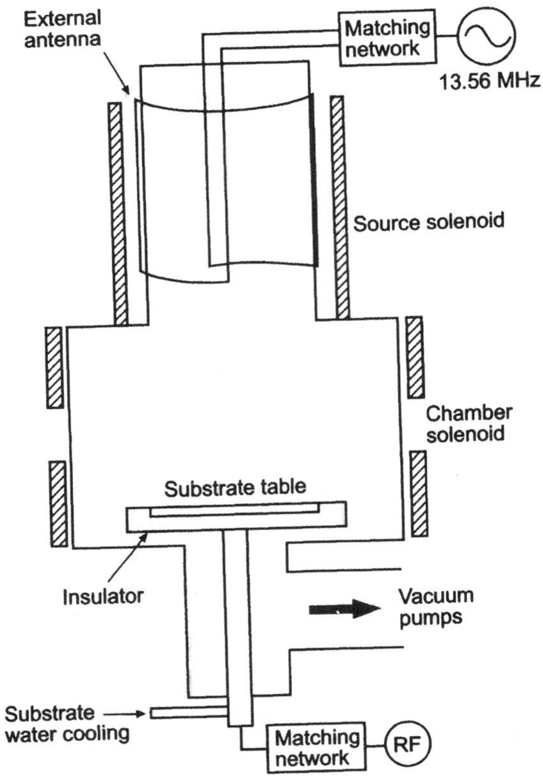

Fig. 4-10 High-density plasma reactor with helicon source (from [15], reprinted with permission).

magnetic field is strong enough, such that the ion cyclotron radius is smaller than the radius of the tube, it can also confine the ions by reducing their loss to the walls. For a tube 15 cm in diameter, a magnetic field of $\sim$ 100 G can achieve the confinement.

The helicon source is operated in the pressure range 10^{-4}–10^{-2} torr, and the plasma diffuses from the source into the reaction chamber. The diffusion is controlled with additional magnets placed around the chamber. These can be permanent magnets in a multipolar configuration or electromagnets [15]. A reaction chamber 30 cm long and 30 cm in diameter can be operated with a helicon source 22 cm long and 15 cm in diameter. The helicon source can be operated over a large frequency range, the range 2–70 MHz appearing to be most suitable [15].

The RF reactors in the different configurations are the most versatile reactors and are most used for plasma processing. The RF reactors can process electrically conductive or insulating samples and are used for deposition of inorganic or polymeric films, for surface treatment of organic or inorganic solids, and for plasma and reactive ion etching.

4.4 MICROWAVE REACTORS

The microwave plasmas are sustained by coupling the microwave energy to the plasma gas through applicators of various designs. Microwaves are easily absorbed or reflected by most materials and cannot be transmitted via cables, like RF energy, without significant losses. Special coaxial cables can be used to transmit low-power (< 200 W) microwaves. The coax consists of two concentric conductors separated by a dielectric. For higher-power levels, specially designed waveguides have to be used. The waveguides are hollow rectangular tubes of high electrical conductivity. The cross sections of the conductors of the coax or the waveguide are determined by the wavelength of the microwave [16].

A microwave power supply system used for microwave or ECR plasma generation is illustrated in Fig. 4-11 and consists of

1. A filtered, low-ripple microwave power supply of constant frequency but variable power.
2. A circulator, whose role is to protect the power supply from large reflected power which may result from an impedance unmatched microwave applicator.
3. Meters for monitoring both the incident and reflected power.
4. A variable, manual or automatic, impedance matching of the applicator with plasma, at different and variable discharge conditions; an impedance mismatch will cause the microwave to reflect instead of propagate into the plasma.

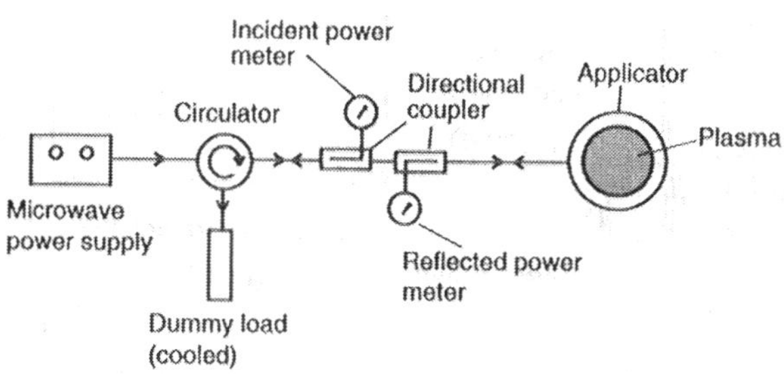

Fig. 4-11 Diagram of microwave power system for cold plasma excitation.

The microwave applicators can be of several types: waveguide, cavity, and coaxial.

The simplest way to transfer microwave power to the plasma is by inserting a dielectric tube reactor through a rectangular waveguide at the center of its wide face, as illustrated in Fig. 4-12. The axis of the reactor coincides with the position

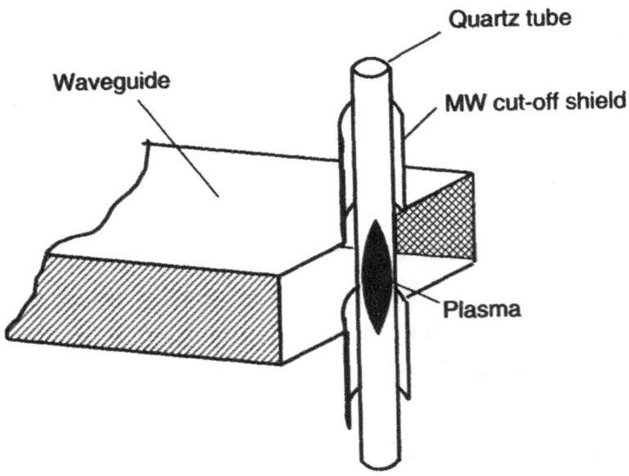

Fig. 4-12 Waveguide-tube microwave coupling.

of the highest electric field in the waveguide, facilitating the acceleration of the electrons by the microwave. To prevent leakage of microwave outside the reactor, the discharge region is enclosed on both sides by metallic tubes with diameters smaller than the cut-off diameter corresponding to the wavelength of the used microwave [16].

The diameter of the reactor is limited in this case by the size of the waveguide, whose width for a frequency of 2.45 GHz is 7.21 cm.

Fig. 4-13 illustrates a different way of microwave coupling to the plasma by a quarter-wave microwave cavity or an Emerson antenna. It is essentially a resonance cavity that is fed by a coaxial cable and is suitable for low powers (< 200 W). The cavity has a removable cap that allows easy disconnection of the

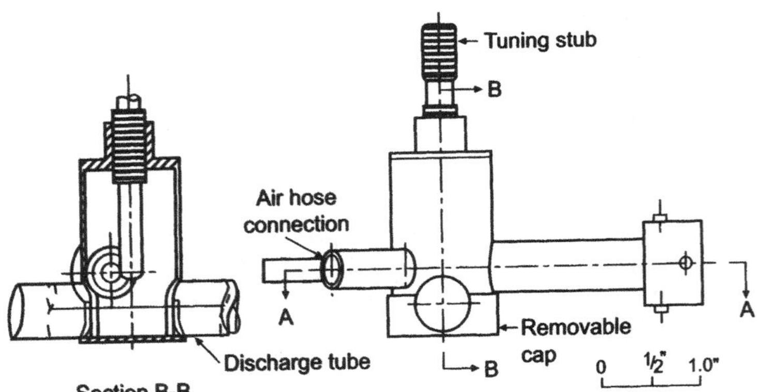

Fig. 4-13 Coaxial $\lambda/4$ cavity (from [17], reprinted with permission from *Plasma Chemistry in Electrical Discharges*, p. 65, 1967).

reactor from the cavity. For microwaves of a frequency of 2.45 GHz, the diameter of the reactor is generally limited to less than 2 cm.

Microwave plasmas of larger volumes than those attainable by direct coupling through the waveguide (Fig. 4-12) can be produced by coupling the microwaves through a larger cavity. Special axisymmetric microwave couplers can convert the microwaves from the rectangular waveguides to circular applicators. In a circular applicator the plasma is symmetrical and can have a nearly uniform radial electric field profile. An example of an axisymmetric applicator that produces plasmas for processing wafers 3–4 inches in diameter [18] is shown in Fig. 4-14.

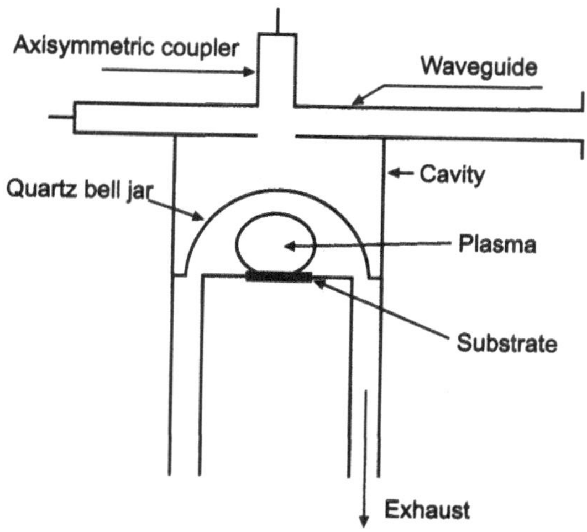

Fig. 4-14 An axisymmetric microwave coupler (from [18], reprinted with permission from *Research & Development Magazine*, October, 1989, by Cahners Publishing Company).

Microwave reactors have been designed for plasma treatment of even larger areas. A reactor designed for large-area etching of polymers [19, 20] combines a long and narrow microwave applicator (42×7.5 cm^2) with a translational movement of the substrate, in a direction perpendicular to the applicator.

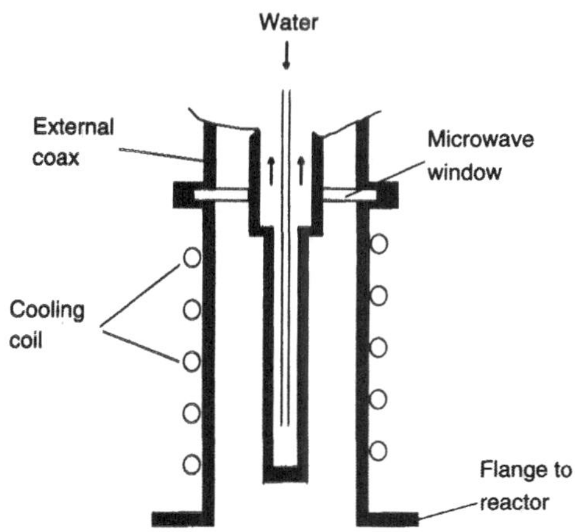

Fig. 4-15 Coaxial microwave coupling to a plasma reactor.

Microwave power can be coupled into a plasma also by using coaxial applicators. As illustrated in Fig. 4-15, the central conductor of the coax is in direct contact with the plasma and has to be generally water cooled. The insulator that separates the central conductor from the external one has also to provide a vacuum seal at the insertion in the reactor.

4.5 ECR REACTORS

Several configurations have been developed for ECR sources for cold plasma reactors. A typical system of the ***divergent magnetic field*** type [21] is illustrated in Fig. 4-16. In this system, microwave power is fed by waveguide into the ECR cavity, through a quartz or alumina window. Solenoidal electromagnets (see Fig. 4-16) create a magnetic field that reaches the ECR value of 875 gauss in the source and diverges outside the source. Electrons diffuse faster than ions into the

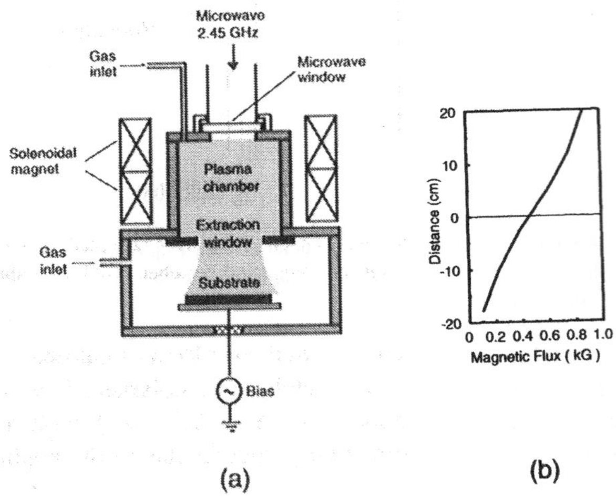

Fig. 4-16 Diverging field–type ECR reactor: (a) plasma source and reactor; (b) distribution of magnetic field along the plasma.

processing chamber, along the diverging magnetic field lines, creating an electric field which causes the extraction of the ions. A potential of 15–20 eV is commonly developed and is imparted to the ions extracted from the source. If higher ion energy is needed, it can be achieved by placing extracting grids between the ECR source and the processing chamber. The grids can also improve the uniformity of the extracted plasma beam.

Additional magnets are sometimes added outside the source to permit shaping of the magnetic field in the vicinity of the substrate and to achieve better control and confinement of the plasma. The auxiliary magnets can be located behind the substrate holder to improve the uniformity of ion current density across the substrate and to ensure the arrival of ions perpendicular to the surface.

The normal ion incidence on the substrate is especially important in etching where ion bombardment contributes to etching anisotropy. In film deposition, the normal arrival of the ion on the coated surface is essential for preventing void formation when filling features of high-aspect ratio [22]. If additional substrate bias is required, it can be achieved by connecting a RF power supply to the sample holder as indicated in Fig. 4-16.

Another type of ECR plasma system is the *multipolar ECR*, or *distributed ECR*, reactor [23] illustrated in Fig. 4-17. In this system, the microwave power is fed through several internally cooled microwave antenna rods, located inside the periphery of the reactor and close to its wall. Permanent samarium cobalt magnets positioned outside the reactor and opposite the antennas create resonant field cusps next to the antennas, as shown in Fig. 4-18. In this type of reactor the plasma diffuses out of the ECR regions, creating a uniform plasma inside the reactor. The internal plasma region is practically free of magnetic fields. The configuration of the magnetic field also slows the loss of particles to the wall and helps maintain a high plasma density. The type of reactor described here is also named *multicusp geometry* reactor.

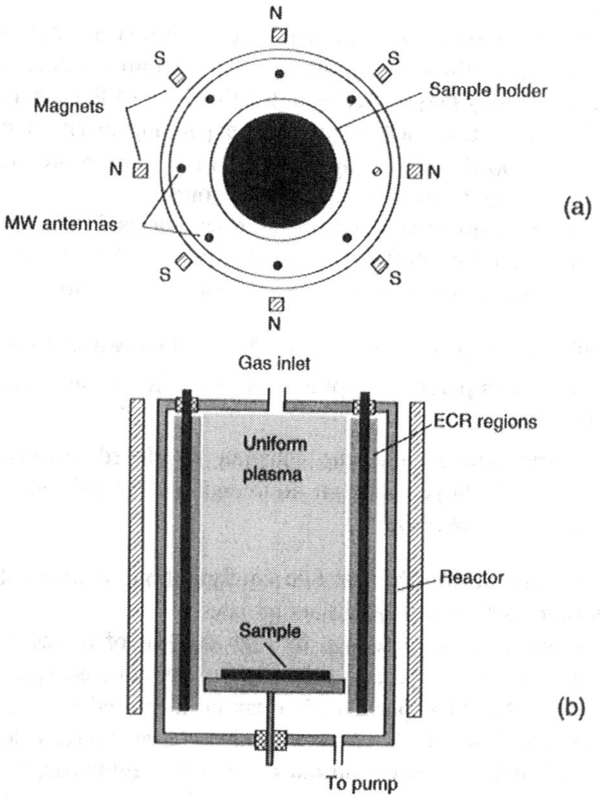

Fig. 4-17 Multipolar ECR plasma reactor.

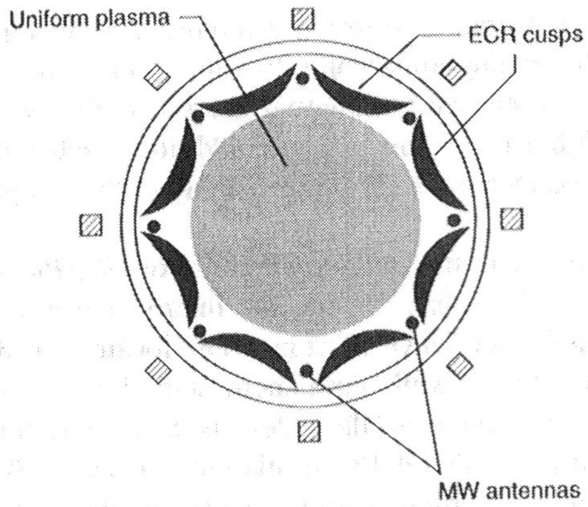

Fig. 4-18 ECR region in multipolar ECR reactor.

In the diverging field ECR plasma system shown in Fig. 4-16, the ions flow along the magnetic field lines, making it a line-of-sight process, while in the multipolar ECR the plasma is essentially isotropic in the internal processing region of the reactor.

For some applications, the substrates have to be treated while having a linear translational movement through the plasma. This requires a linear plasma source having a uniform density perpendicular to the direction of the substrate movement. A special ECR source using a horn antenna expanding to an aperture of 60×9 cm^2 has been developed for such applications [24]. The source is called by the authors an *elongated-mirror-geometry* (*EMG*) source.

Because of the short wavelength of the microwaves, both regular microwave and ECR plasmas can be confined to small volumes. As a result, processing in microwave/ECR plasmas can be performed in several reactor arrangements:

- The substrate is placed entirely inside the microwave/ECR discharge.
- The substrate is placed in a processing chamber, located downstream from the discharge.
- In a hybrid arrangement, the substrate is placed downstream from the microwave discharge, with an additional RF or DC discharge sustained adjacent to the substrate.

Processing performed using the last two configurations is also called *processing by remote plasma*, which will be discussed later.

The ECR plasmas, characterized by high degrees of ionization, can produce high-density plasmas over a broad range of pressures. The energies of the ions are intrinsically low in the ECR plasmas, but can be increased using extraction grids. Because of these features, ECR reactors produce large fluxes of low-energy ions, are characterized by high processing rates, and are therefore used for single-wafer processing, for etching of wafers, or for deposition of coatings. The large mean free path of the particles at the low operating pressures of the ECR plasmas results in directionality of the ion and neutral beams. When used for anisotropic etching, ECR reactors cause lower wafer damage than reactive ion etching performed in RF reactors at much higher ion energies.

4.6 MAGNETICALLY ENHANCED REACTORS

The high-density plasmas obtained in the ECR reactors are a result of the interaction between the electric field at microwave frequencies and the superposed magnetic field. The value of the magnetic field has to be adjusted to create resonance between the frequency of the electric field and the electron cyclotron frequency. Magnetic fields can be used to enhance plasma densities also at nonresonant conditions. Two types of magnetically enhanced plasma reactors are described in the following.

The *magnetically confined reactor* (*MCR*), is essentially a triode reactor that uses additional magnetic fields to confine the plasma and to reduce the loss of

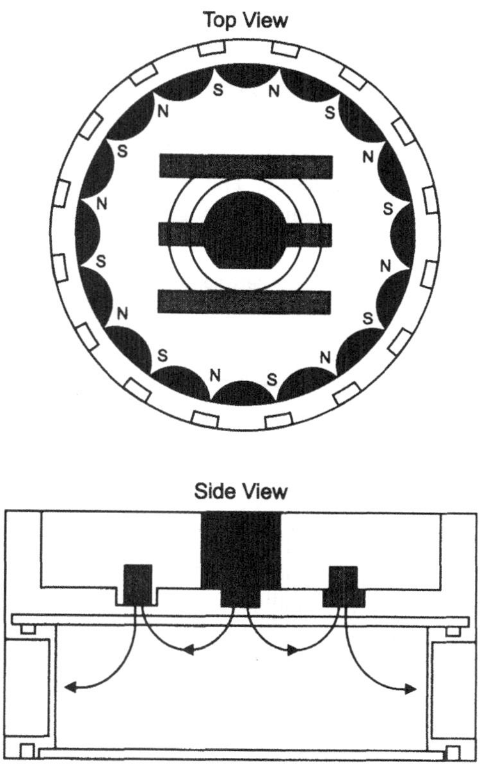

Fig. 4-19 Magnet arrangement in a magnetically confined reactor. The arrows indicate the magnetic field lines (from [25], reprinted with permission).

charged species to the walls. The triode reactor, similar to the one shown in Fig. 4-8, uses two power supplies at different frequencies: one for exciting the plasma and one for biasing the substrate electrode. A 13.56-MHz power supply is used to sustain the plasma, while a 100-kHz power supply provides the bias to the substrate. The magnetic confinement is produced by permanent magnets arranged around the perimeter of the reactor and imbedded in the grounded upper electrode, as illustrated in Fig. 4-19.

The power supplied by the high-frequency field controls the amount of dissociation and generation of reactive species. The low-frequency field controls the ion flux to the substrate and enhances the directionality of the process, especially in etching.

The confinement of the plasma by the magnetic fields results in an enhanced plasma density. The arrangement of the magnets keeps the magnetic filed lines away from the processed wafer (see Fig. 4-19), preventing a direct influence of the magnetic field on the process. The MCR is used at pressures of a few mtorr [25, 26]. Being based on RF power supplies and using a magnetic field that is independent from the frequency of the electric field, the equipment used for the MCR is much simpler than that required for ECR reactors.

A different version of a magnetically enhanced plasma reactor used especially for magnetically enhanced reactive ion etching (MERIE), is the *rotating field reactor*. This is essentially a parallel plate RF reactor in which a time-varying magnetic field is applied on the substrate electrode. As shown in Fig. 4-20, several electromagnets are arranged around the substrate electrode, which in most cases serves as the cathode of the discharge. The plasma is sustained by a 13.56-MHz generator. The electromagnets produce a magnetic field that rotates slowly, at a

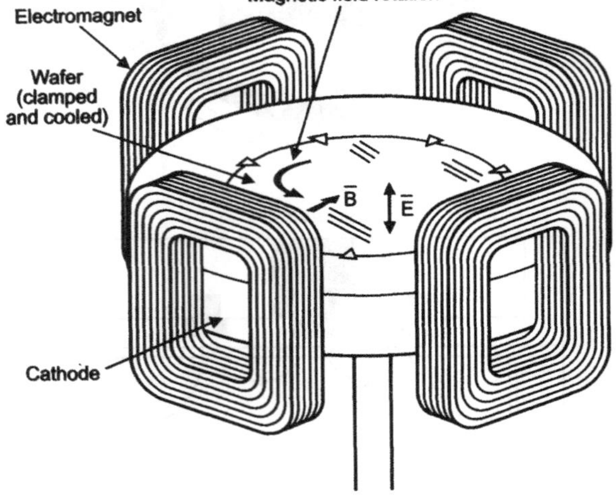

Fig. 4-20 Cathode and electromagnets in a rotating field reactor (from [27], reprinted with permission).

frequency of 0.5 Hz, in the plane of the electrode. This arrangement produces a high flux of low-energy ions, resulting in high processing rates with reduced substrate damage [27]. Both MCR and rotating field reactors are considered alternatives to ECR reactors, especially for plasma etching.

4.7 REMOTE PECVD REACTORS

In the reactors previously discussed, the processed samples are inserted directly in the plasma excited in the feed gases. The processing in such reactors is sometimes called processing by *direct PECVD*. A different type of cold plasma processing is by *remote PECVD* [28]. The remote PECVD is also called *downstream processing*.

The distinction between direct and remote PECVD is made primarily on the basis of whether reactant or diluent gases are directly excited by the plasma. In the direct processes described before, all process gases are exposed to the plasma. In addition, the processed samples are immersed in the plasma discharge.

By contrast, in the remote PECVD processes, not all reactant gases are excited in the plasma, and the substrate is removed from the discharge region of the plasma. The gases can be introduced into both the discharge region and the reaction chamber outside the discharge, often near the substrates. This arrangement reduces the number of possible reaction pathways and can improve control of the process or the stoichiometry of deposited films [28]. Physical effects of the plasma on the processed substrate can be reduced in the remote reactors, minimizing or completely eliminating radiation damage.

A remote plasma reactor is illustrated in Fig. 4-21. In such a system some of the process gases (dilutants and/or reactants) are excited in the plasma sustained in a quartz tube by inductive RF coupling, while other reactant gases are introduced in the processing chamber through a dispersion ring. The processed substrate is supported outside the plasma.

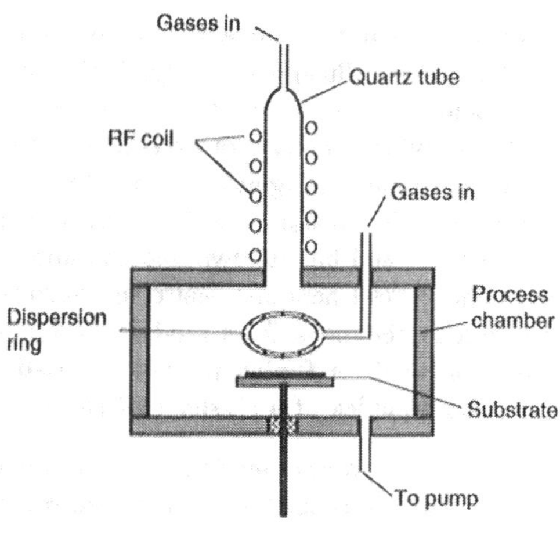

Fig. 4-21 Reactor chamber and RF plasma source for remote PECVD.

Hybrid systems combine characteristics of both direct and remote processes. This is the case in some RF triode-type reactors in which the substrate sits below a grid that replaces one of the electrodes and it is not exposed directly to the discharge. A regular ECR reactor can also be considered a hybrid processing system because the substrate is generally outside the plasma chamber while all the gases may be excited in the plasma.

An ECR reactor can also be used as a remote PECVD reactor. The gases can be introduced both in the ECR plasma chamber and in the processing chamber, and as mentioned earlier, the substrate can be placed outside the plasma. For example, to deposit silicon dioxide by the remote process, oxygen is introduced into the system via the plasma chamber (the ECR source), while silane enters directly in the reaction chamber. The oxide is formed by the reaction that takes place on the substrate surface between a silane molecule and incident oxygen ions. For deposition of silicon nitride the arrangement is similar, but in this case the plasma gas is changed to nitrogen [22].

Remote plasma reactors are used when it is necessary to avoid direct contact of the processed sample with the plasmas, to prevent damaging of sensitive samples. They are used for deposition of inorganic or organic coatings or for plasma treatment of polymers.

4.8 REACTOR CLUSTERS

Many manufacturing processes using PECVD consist of multiple steps, with one or more plasma reactors. This is especially the case in manufacturing micro-electronic components. For example, manufacturing of DRAM computer chips involves hundreds of process steps and the processing of the 64-Mbit DRAM chips will require up to 1000 steps [29].

For better control of the fabrication process and yield improvement, the wafers have to be moved between process steps without exposure to environment to avoid contamination. For this purpose, cluster tools that are modular, multi-chamber, integrated process systems were developed in the last few years. A cluster tool, as illustrated in Fig. 4-22, consists of a central handling station (the core) and a number of peripheral modules also called process stations. Processing modules, which include the reactors, load-locks, or inspection stations, can be arranged in any configuration around the central core. PECVD reactors are often included in the cluster tools. Each processing unit is mechanically and electrically independent and has its own vacuum and control system.

The central handling platform contains transport mechanisms that move the processed specimens from module to module. Several central platforms may be interconnected for further process integration.

The modules of a cluster tool can be of two types:

- Single-process modules, which are usually batch modules for processing multiple specimens in low rate processes.

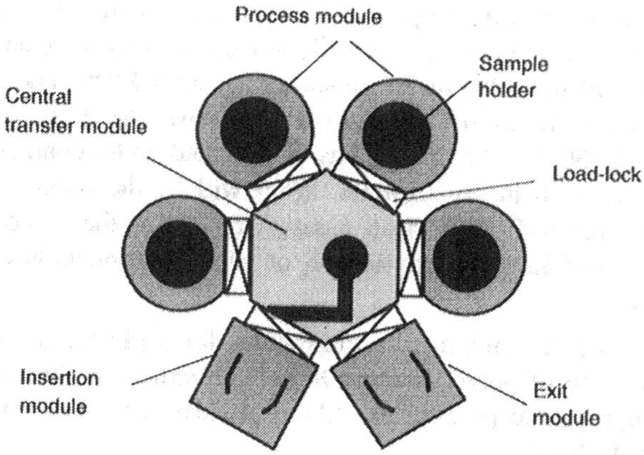

Fig. 4-22 Diagram of a cluster tool.

- Multiple-process modules, which consist of several fully separated internal stations, each providing a single-wafer processing environment. A multi-process module occupies only a single port of the central handling module. An internal transport mechanism moves the wafers from one process station to another within the module, while the central handler is free to serve other ports of the cluster.

4.9 QUESTIONS

1. A certain plasma process has an optimum for a residence time of 0.4 sec, when the process is performed in the pressure range of 100 mtorr to 1 torr. The process will be performed in a parallel plate reactor having two electrodes, 8 in. in diameter and 3 in. apart. One electrode serves as a showerhead for introduction of the gases in the system. What is the maximum range of mass flow controllers to be used with this reactor? Explain.
2. The mass flow controller to be used with a plasma reactor requires a minimum pressure difference of 10 psi between its ends for proper operation. The plasma process performed at a pressure of 5 torr in the reactor has to use a precursor that has a vapor pressure of 600 torr at room temperature. Describe an experimental arrangement for supplying the precursor to the reactor at controlled flow rate.
3. Another precursor has a vapor pressure of 50 torr at room temperature. Describe the experimental arrangement for supplying this precursor at a controlled flow rate. What is determining the flow rate?
4. One of the precursors for a plasma process has a very-low-vapor pressure at room temperature. What experimental arrangement is required to supply a controlled flow rate of this precursor to the reactor?

5. A plasma process was developed in a cylindrical reactor, 10 cm in diameter, in which the plasma is inductively excited by a 7-cm-long coil. The coil is powered with 50 watts by an RF generator at 13.56 MHz. The process has an optimum at a pressure of 0.1 torr, a total mass flow rate of 10 sccm and sample temperature of 150 °C. The process has to be scaled up to a parallel plate reactor with electrodes of 8 in. in diameter. What will be the process parameters in the scaled-up reactor? Will these parameters ensure the reproduction of the process developed in the small reactor, or will adjustments have to be made? Explain why.
6. Describe the special equipment requirements for a plasma system to be used for the deposition of polycrystalline films with extremely low oxygen content employing pyrophoric precursors. Which process parameters will have to be precisely controlled?
7. Which type of reactor appears to be most suitable for high rate reactive etching with minimal induced substrate damage?
8. List all hardware components required to set up a diverging field ECR reactor. What changes will be required for a multipolar reactor?
9. Design a parallel plate triode reactor and list the hardware components required for this reactor. Describe different equipment options.

4.10 REFERENCES

[1] Hollahan, J. R., and R. S. Rosler, In *Thin Film Processes*, eds. J. L. Vossen and W. Kern, p. 335. New York: Academic Press, 1978.

[2] Vossen, J. L., and W. Kern, eds., *Thin Film Processes*. New York: Academic Press, 1978.

[3] Catherine, Y., In *Plasma Processing*, eds. G. S. Mathad, G. C. Schwartz, and G. Smolinsky, p. 317. Pennington, NJ: Electrochemical Society, 1985.

[4] Hess, D. W., and D. B. Graves, In *Microelectronics Processing*, eds. Dennis W. Hess and Klavs F. Jensen, p. 377. Washington, DC: American Chemical Society, 1989.

[5] Mort, J., and F. Jansen, eds. *Plasma Deposited Thin Films*. Boca Raton, FL: CRC Press, 1986.

[6] Yasuda, H. K., *Plasma Polymerization*. New York: Academic Press, 1985.

[7] U.S. Federal Communications Commission Rules and Regulations. Washington, DC: U.S. Government Printing Office, 1964.

[8] Mayo, N., U. Carmi, I. Rosenthal, R. Avni, R. R. Manory, and A. Grill, *J. Appl. Phys.*, 55: 4404 (1984).

[9] Grill, A., A. Raveh, and R. Avni, *Surf. Coat. Technol.*, 43/44: 745 (1990).

[10] Reinberg, A. R., *U.S. Patent*, 3, 757, 733 (1973).

[11] Reinberg, A. R., In *Proc. Electrochem. Soc. Meet.,* p. 30. San Francisco, May 12-75, 1975.

[12] Sherman, A., *Thin Solid Films*, 113: 135 (1984).

[13] Chapman, I. B., *Glow Discharge Processes: Sputtering and Plasma Etching.* New York: J. Wiley & Sons, 1980.

[14] Flamm, D. L., *Solid State Technol.*, 34(3): 47 (1991).

[15] Perry, A. J., D. Vender, and R. W. Boswell, *J. Vac. Sci. Technol.* B9, 310 (1991).

[16] Marcuwitz, N., *Waveguide Handbook.* London: Peter Peregrinus, 1986.

[17] McTaggart, F. K., *Plasma Chemistry in Electrical Discharges.* Amsterdam: Elsevier, 1967.

[18] Post, R. S., J. R. Conrad, and D. K. Smith, *Res. Dev. Mag.*, 31(10), 106 (1989).

[19] Lamontagne, B., A. M. Wrobel, G. Jalbert, and M. R. Wertheimer, *J. Phys. D.*, 20: 844 (1987).

[20] Wrobel, A. M., B. Lamontagne, and M. R. Wertheimer, *Plasma Chem. Plasma Process.* 8: 315 (1988).

[21] Matsuo, S., and M. Kiuchi, *J. Appl. Phys.* 22: L210 (1983).

[22] Denison, D. R., M. S. Chang, and T. Ebata, *Microelectron. Manufact. Test.* 22 (July 1990).

[23] Burke R. R., and C. Pomot, *Solid State Technol.* 31(2): 67 (1988).

[24] Geisler, M., J. Kieser, E. Rauche, and R. Wilhelm, *J. Vac. Sci. Technol.* A8: 908 (1990).

[25] Engelhardt, M., *Semicond. Int.*, 52 (July 1991).

[26] Laporte, P., Y. Melaku, and M. Goethals, *Solid State Technol.* 34(4): 69 (1991).

[27] Nguyen, S. V., G. Christman, D. Dobuzinsky, and D. Harmon, *Solid State Technol.*, 33(10): 73 (1990).

[28] Lucovsky, G., D. V. Tsu, and R. J. Markunas, In *Handbook of Plasma Processing Technology*, eds. Stephen M. Rossnagel, Jerome J. Cuomo, and William D. Westwood, p. 387. Park Ridge, NJ: Noyes Publications, 1990.

[29] Bader, M. E., R. P. Hall, and G. Strasser, *Solid State* Technol. 33(5): 149 (1990).

5

Plasma Diagnostics

In the plasma, a large number of variables affect the balance between competing chemical and physical processes. The number of parameters is normally so big that the use of a random sampling technique to achieve optimization of reactor conditions for a given process would be extremely cumbersome. Appropriate reliable diagnostic tools are therefore necessary to provide an understanding of the plasmas and the actual phenomena taking place, which in turn will enable the control of the process.

The objective of plasma diagnostics is to obtain information about the state of the plasma by means of experimental examinations of physical processes occurring in it. Knowledge of plasma physics is required to understand fully the effects of the physical processes taking place in the plasma and to deduce from them its properties. The discussion of plasma physics is, however, beyond the purpose of the present book, and only the basic principles, those related to the specific diagnostic tools, will be presented.

The diagnostic techniques can be classified into *ex-situ* and *in-situ techniques*. The ex-situ techniques sample the contents of the plasma reactor and transfer a fragment of the contents outside the reactor for examination. Among ex-situ or *off-line* techniques that are used for plasma diagnostics are mass spectrometry and gas-phase electron paramagnetic resonance. Since the contents of a plasma reactor include many highly reactive species, such as ions, radicals, and reactive atoms, one can never be certain that species analyzed ex-situ have not undergone changes during their transfer from the reactor.

The in-situ or *on-line* techniques can be further classified into *intrusive* and *nonintrusive* methods, although this division is somewhat arbitrary. Any diagnostic technique is to some extent intrusive and perturbs the plasma. In some cases, this perturbation may be negligible, while in others it may be quite significant. The degree to which the perturbation is significant depends on the required information.

The diagnostic methods are also sometimes classified according to the parameter of the plasma to be determined or by the experimental technique used. As the first approach has the drawback that many methods typically measure more than one parameter, the second approach will be adopted in this chapter.

The most used techniques for plasma diagnostics, namely, mass spectrometry, electric probes, and optical methods, will be explained in this chapter. More detailed discussion and analysis of the different diagnostic methods can be found in specialized treatises on plasma diagnostics, such as Huddlestone and Leonard [1], Loechte-Holtgreven [2], Venugopalan [3], Heald and Wharton [4], Hutchinson [5], Auciello and Flam [6]. Applications of different plasma diagnostic tools for process characterization can be found in many papers dealing with different processes and in various reviews, such as, for example, [7, 8].

5.1 MASS SPECTROMETRY

Mass spectrometry is one of the most used tools for diagnostics of cold plasmas in research laboratories as well as in manufacturing plants. The mass spectrometer is a device of high detection sensitivity, which allows the identification of molecules, radicals, atoms, or ions by their molecular or atomic weight.

The mass spectra of gaseous components in cold plasmas are relatively simple. As a result, mass spectroscopy is a very useful method for quantifying the composition of the plasmas in the study of plasma chemistry or in controlling a manufacturing process.

Mass spectrometry can be used either for identification or for quantitative analysis of plasma species, and the results can be used for the determination of the kinetics of the reactions in the plasma or for controlling the process to obtain the desired results.

Several types of mass spectrometers are available: the magnetic analyzer, which selects ions of similar momentum; the time-of-flight spectrometer, which analyzes ion velocity; and the quadrupole mass spectrometer, which analyzes ions on the basis of their mass to charge, m/e, ratio (atomic mass units, amu). In recent years most mass spectrometry studies of cold plasmas have been performed using the quadrupole *mass spectrometer* (*QMS*), first described by Paul et al [9].

5.1.1 The Quadrupole Mass Spectrometer

A state-of-the-art discussion of the quadrupole mass spectrometer, its theory and operation, were given by Dawson [10]. Detailed descriptions of applications of mass spectrometry for plasma diagnostics can be found in reviews by Venugopalan [3], Austin et al. [11], and Vasile and Dylla [12]. The quadrupole mass spectrometer and its application for plasma diagnostics are explained in the following sections.

5.1.1.1 The Quadrupole Mass Analyzer

The main part of the quadrupole mass spectrometer is the *quadrupole mass analyzer*, or *mass filter*. A quadrupole consists of four conducting rods or poles, the axes of which are placed at the corners of a square, as shown schematically in Fig. 5-1. The poles often have hyperbolical surfaces and typical dimensions of the analyzer are

$$\text{pole length} = 16 \text{ cm} \qquad r = 0.3 \text{ cm}$$

Opposite poles are connected electrically together, and a potential $\Phi(t) = U + V\cos(\omega t)$ is applied to one pair of rods, while the same potential of opposite

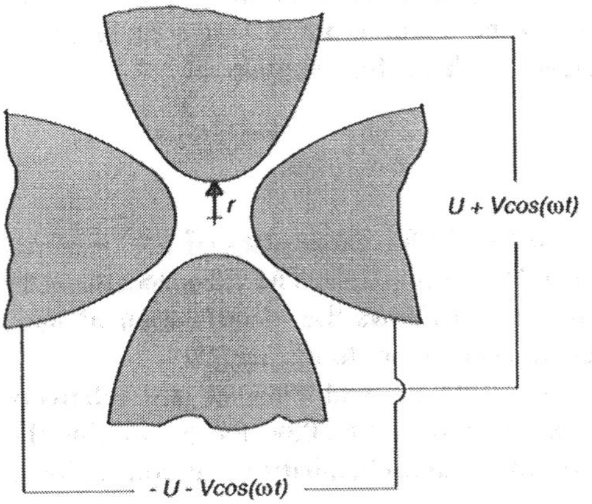

Fig. 5-1 Cross section of the quadrupole mass spectrometer rods with electrical connections.

sign, $-\Phi(t)$, is applied to the other pair. U is a DC voltage and V is the amplitude of a RF voltage of frequency ω. If an ion enters the quadrupole along a direction approximately parallel to the axes of the cylinders (z-axis), the applied potentials will induce a transverse component to the motion of the ion.

The RF field removes low-mass ions from the beam entering the analyzer and focuses high-mass electrons close to the axis of the analyzer. The superimposed DC field removes ions of high mass from the beam. For a certain combination of the parameters, U, V, ω, and m/e, the ion will oscillate along the z-axis and emerge at the second end of the quadrupole. For all other combinations of parameters, the ion will move away from the z-axis, ultimately strike an electrode, and be neutralized. By a suitable selection of V and U values, only ions of a given m/e range will have a stable motion and will be transmitted through the quadrupole. The quadrupole thus acts as a m/e filter that can be tuned to a chosen mass by adjusting the amplitude of the RF voltage.

For fixed U and V values and constant ion charge, there is a range ΔM of masses for which the ions will pass through the quadrupole analyzer. This range

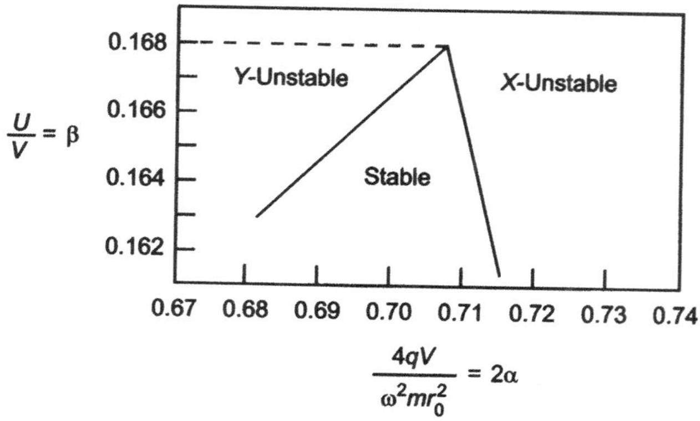

Fig. 5-2 Diagram of stable solutions for the movement of ions through quadrupole mass analyzer (from [13], reprinted with permission).

of masses, having a stable motion along the quadrupole, depends on the ratio $\beta = U/V$ and decreases with increasing β, as illustrated in Fig. 5-2. The ratio between the DC and RF voltages thus determines the *selectivity of the analyzer*.

For $\beta = 0.168$, only particles of one m/e value can pass through the analyzer. The corresponding resolution of the instrument ($R = M/\Delta M$) will be infinite; however, the current of ions arriving to the detector will be infinitesimally small. Values of $\beta < 0.168$ are therefore used to increase the value of ΔM. The range of analyzed masses can be scanned by scanning U and V while keeping β fixed. As can be seen in Fig. 5-2, for fixed β the resolution R of the analyzer is constant. In many cases β is adjusted during scanning to keep ΔM constant instead of R.

The amplitude of the RF voltage, V, and its frequency, ω, determine the mass of the ions that are allowed to pass through the analyzer. By sweeping the RF and DC voltages from zero to their maximum value, while keeping their ratio β constant, ions will be transmitted sequentially in order of their m/e values. The ions, which pass through an aperture at the end of the apparatus, are detected by an ion detector, which can be either a Faraday cup, or an electron multiplier.

A quadrupole mass analyzer is characterized by a *transmission versus mass curve*. The transmission of the instrument is a function of the velocity of the ion, and at a fixed energy, ions of high mass will be transmitted less efficiently than will ions of low masses. The plasma diagnostics requires the determination of the quantitative relationship between species from different groups of masses. The decrease in *sensitivity* of the quadrupole with increasing mass must therefore be kept in mind when performing quadrupole mass spectrometry plasma diagnostics and particularly in the calibration of the instrument as described later [12].

The quadrupole mass analyzer's characteristics prove advantageous for mass spectrometry in general and plasma diagnostics in particular. Among those properties are

- Independence of the mass filtering from the energy distribution of the ion beam.
- High transmission rate due to the continuous focusing along the rods; this results in high sensitivity.
- Fast scanning (the feature can be exploited when a suitable ion detector is used).
- Small dimensions.
- Ease of mounting and alignment to the analyzed system.
- Low overall cost.

An additional advantage of the quadrupole mass analyzer for plasma diagnostics is the possibility of controlling the mass filtering characteristics, that is, resolution versus transmission, by variation of electrical parameters only.

5.1.1.2 The Ion Source

The quadrupole mass spectrometer can be used to analyze both ions and neutrals. To analyze neutral species, such as atoms, molecules, or radicals with the quadrupole mass analyzer, they have to be ionized first. This is achieved in a ionization source placed in front of the quadrupole analyzer. The ionization of the species extracted from the reactor is performed in the ion source by a beam of electrons passing through the vapor that flows through the source. A typical ion source is presented in Fig. 5-3.

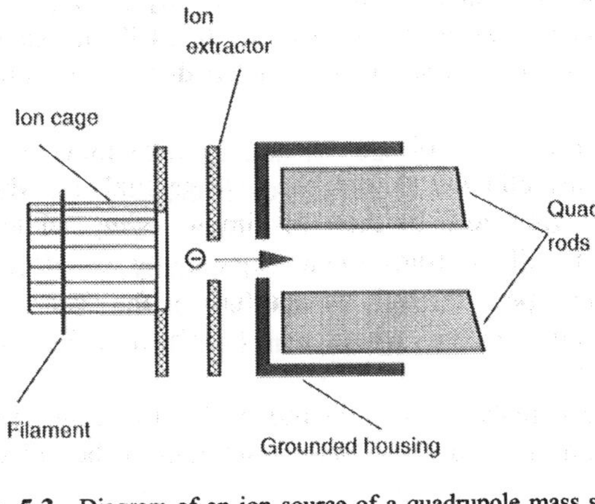

Fig. 5-3 Diagram of an ion source of a quadrupole mass spectrometer.

A heated thermionic filament, usually made of thoria-coated iridium wire, produces an electron beam that is accelerated to the ion cage, which serves as an anode at a potential of about + 70 V DC. The neutral particles, passing normal to the electron beam, are ionized by electron impact. An ion extractor, or ion focus electrode, extracts the ions from the ionization source, focuses them into a narrow beam, and projects them into the quadrupole filter, along its *z*-axis (parallel to the direction of the rods).

Although negative ions are also formed sometimes in the ionization process by the attachment of an electron to a neutral particle, only the positive ions generated in the source are generally used for the analysis of the beams ionized in the source.

5.1.1.3 The Detector

The ions filtered by the quadrupole mass analyzer are collected by a suitable detector. For many applications, a *Faraday plate collector* is sufficient to collect and measure the ion current corresponding to ions of specific m/e value. The signal of the detector is the current required to discharge the ions striking the collector. This current is consequently amplified. Because the sensitivity for nitrogen is 2×10^{-4} amp.torr^{-1}, the current of the Faraday plate (or cup) is very low. The Faraday plate detector is very stable and does not pose a lifetime problem.

For detection limits higher than those achievable with the Faraday detector and for faster response required in rapid mass scanning, electron multipliers are used. A *secondary electron multiplier* (*SEM*) acts as preamplifier of the ion signal. It consists of several stages, also called dynodes, with surfaces characterized by a high yield of secondary electron emission. The first dynode operates at about − 2000 V. An energetic ion accelerated at that voltage, striking the sensitive surface (often made from CuBe alloy), causes the emission of a number of secondary electrons, which are accelerated to the next stage of the detector, where in turn they cause emission of secondary electrons. The acceleration of the electrons is performed at a voltage sufficiently high to produce a secondary electron yield much bigger than unity, so that the number of electrons is multiplied at each stage. The number of stages in the electron multiplier, about 15, is limited by the maximum current carrying capacity of the latest stage. The gain of the electron multiplier depends strongly on the operating voltage and the condition of the surfaces of the dynodes, that is, whether fresh or oxidized. Typical gain of an electron multiplier operating at − 2000 V is about 10^5.

The sensitivity of the CuBe surface of the dynodes is affected by exposure to air or by hydrocarbon contamination. Exposure of the electron multiplier to air should therefore be avoided as much as possible. A main source of hydrocarbon contamination is the backstreaming of oil vapors from vacuum pumps. This kind of contamination can be eliminated by using turbomolecular pumps. The deterioration of the dynode surfaces can result in a reduction of the gain by 50 to 90%. To compensate for this deterioration, the electron multiplier can be operated over a range of voltages, for example, 1000 to 3000 V, to adjust its gain which generally changes with time.

The sensitivity of Faraday cup detectors is not dependent on the mass of the ion; that is, the Faraday detector has no mass discrimination. Secondary electron multipliers may have mass discrimination, depending on their design and operating voltage. At high voltages their sensitivity is higher for ions of larger mass.

The electron multiplier has the disadvantage that its gain tends to vary with time. Therefore, the electron multiplier requires regular calibration. The Faraday plate detector is more accurate and simpler to operate. The applicability of the different detectors is dependent on two characteristics, namely, the detection limit and response time to be discussed next.

Detection Limits

Ion currents as low as 10^{-15} amp can be detected with a time constant of the order of 1 sec. Because the sensitivity for masses up to 200 amu is about 10^{-4} amp.torr^{-1}, partial pressures as low as 10^{-11} torr can be detected at the relatively large time constant. As faster response times are normally required, the practical sensitivities are lower, decreasing with decreasing response time.

The Faraday plate detector is characterized by a detection limit as low as 10^{-10} torr. Higher sensitivity can be obtained with the electron multiplier. When potentials between 2 and 4 kV are applied to the secondary electron multiplier, gains between 10^4 and 10^8 can be obtained, corresponding to a detection limit of about 10^{-14} torr.

The secondary electron multiplier ion detector is, however, practically used with the quadrupole mass analyzer in the pressure range of 10^{-11}–10^{-4} torr. The lowest practical measurable pressure represents the detection limit defined by a signal-to-noise ratio (or noise level) of approximately unity after integration of the minimal output current (10^{-11}–10^{-12} amp) of a typical 14-stage CuBe multiplier over a period of 1 sec [12].

The high-pressure limit of about 10^{-4} torr for both types of detectors is usually defined as the pressure at which gas scattering in the ion source becomes significant and the signals become nonlinear with total pressure. Practical sensitivities achievable with a Faraday cup detector and with a secondary electron multiplier are shown in Table 5-1.

TABLE 5-1 Sensitivities of Quadrupole Mass Spectrometer for Determination of Pressure with 10% Accuracy (from [10], reprinted with permission from *Quadrupole Mass Spectrometry and Its Applications*, p. 138, 1976)

Pressure (torr)	10^{-8}	10^{-9}	10^{-10}	10^{-11}
Current (amp)	10^{-12}	10^{-13}	10^{-14}	10^{-15}
Time electron multiplier (sec)	10^{-5}	10^{-4}	10^{-3}	10^{-2}
Time Faraday cup (sec)	0.02	0.2	2	

Although the numbers in Table 5-1 show that the quadrupole mass spectrometer has a dynamic range of 10^7, in practice, this range is rarely achieved in plasma diagnostics because of high background levels. The background is of special concern in plasmas, due to the high reactivity of the plasma species. Especially high backgrounds are often encountered in cold plasmas of practical interest at atomic mass units (amu) of 28 and 29. Possible sources of these backgrounds are

$$CO,\ C_2H_4,\ N_2,\ {}^{28}Si \quad \text{at } m = 28 \text{ amu}$$

and

$$^{13}CO,\ {}^{14}N^{15}N,\ C_2H_5,\ {}^{29}Si \quad \text{at } m = 29 \text{ amu}$$

The lower detection limit mentioned for the secondary electron multiplier refers to molecular species. For free radicals, however, the lower detection limit is higher, and the detection sensitivity is lower by several orders of magnitude due to the rapid destruction of the active species within the system and interference effects from the background. Nevertheless, the sensitivity of the quadrupole mass spectrometry technique is very high, much more than attainable by other methods.

Response Time

The response time of the ion detector is important because rapid mass scans are frequently required. The response of the Faraday plate detector depends on the used amplification and is a function of the allowable noise level. For the lowest noise levels of about 10^{-5}, the response time is several seconds, while the response can be made faster if a higher noise level is allowed. The response time can be lowered to the order of tens of milliseconds at the expense of the detection limit.

The response time of the electron multiplier, which is determined by the motion of electrons in vacuum, can be much faster, on the order of microseconds. The time resolution of the secondary electron multiplier is generally in the range of 30 to 300 kHz.

5.1.1.4 Resolution of the QMS

The *resolving power* of a spectrometer is defined as its ability to separate adjacent mass peaks. The commonly accepted definition for the resolving power of the mass spectrometer is the "10% valley" definition, although a 50% valley definition is also used. According to this definition, the resolving power is the largest atomic mass number at which peaks of adjacent molecular weight have a valley between them, lower than 10% of the height of the peak.

The resolution of a peak of mass M is defined by the ratio $M/\Delta M$, where ΔM is the distance between the points at 10% of the maximum height of the peak of mass M.

Operating the quadrupole mass filter in the constant resolution mode over the entire mass range will cause a decrease in sensitivity for the more massive, lower-velocity ions [11]. Operation in the constant ΔM mode, where the resolution increases with the mass, compounds the effect of the instrument transmission. It is convenient to keep the ΔM constant at a value that ensures adequate separation of masses that are 1 amu apart. In this scanning mode, called constant ΔM mode, the resolution is proportional to M. If $\Delta M = 1$, the resolution required to separate adjacent peaks at mass 30 is 30, while at mass 250 the resolution must increase to 250 to separate the peaks. The result is a mass spectrum with sensitivity skewed toward the lower masses.

In practice, most quadrupole systems can be electrically tuned to operate either at constant resolution mode or at constant ΔM mode. Once a mode has been selected appropriate to the experimental purpose, calibration has to be done with known compounds or standard gas mixtures. The calibration should include peak intensity measurements as well as resolution measurements over the mass range to be investigated.

When the quadrupole analyzer is tuned to detect ions of low masses, both U and V approach zero values. The ratio V/U can become very large, and the analyzer stops acting as a mass filter, letting a large number of ions of unseparated masses to pass through it. A high current called *zero blast* is detected. The zero blast interferes with the detection of ions of masses 1 and 2 by the quadrupole mass spectrometer. Hydrogen is therefore not analyzed with QMS.

5.1.1.5 Calibration

The basic calibration procedure is the establishment of the signal level for each particular gas in the reactor as a function of its partial pressure. This is achieved by adding each source/reactant gas of interest separately to the reactor and measuring the signal for a principal ion in the cracking pattern of this gas in the ion source. The procedure has to be repeated over a range of pressures and/or flow rates, and without a discharge in the reactor. The calibration has to be performed also as a function of the ionization potential in the ion source. The procedure should also include for reference an inert carrier gas such as argon or neon.

Once the correlation between the chosen signal and partial pressure of the gas has been established, the partial pressures of plasma species, corresponding to the precursor gases, can be deduced from the calibrated mass numbers. This is true, provided that there are no plasma species that could cause interference at the calibrated mass numbers. A self-consistency check of the analysis is possible because the measured partial pressures have to satisfy the following equation:

$$\sum_i p_i = p \tag{5.1}$$

where p_i is the partial pressure of the gas i and the summation is performed for all plasma species. The total pressure in the plasma reactor, p, has to be measured independently and compared to the sum of Eq. (5.1).

The identification and measurement of plasma products are much more complicated. They have to be performed by iterative procedures, since the calibration for many of the species produced by the plasma may not be available a priori. The mass spectrum observed for the plasma products has to be interpreted first in terms of possible stable neutral products, which are able to form in the plasma from the starting gas or gas mixture. Additional calibration measurements with molecules that are deduced as products have to be then performed for determination of both the sensitivity factors and cracking pattern at the relevant electron energy in the ion source.

In some cases, the discharge products cannot be obtained as stable molecules. Indirect methods for the determination of sensitivity factors must then be used [14].

The use of this calibration procedure in the analysis of neutrals is not widespread, presumably because of the convenience of using relative methods. There are two disadvantages to adhering to the calibration procedures just outlined, which have apparently discouraged their widespread use in the past, namely, the large volume of calibration data needed for both qualitative analysis (determination of cracking patterns and appearance potentials) and for quantitative analysis (determination of absolute sensitivities, i.e., the ratio between the signal strength and partial pressure). The widespread use of computers for rapid data acquisition makes this tedious task far less forbidding than it was before.

5.1.2 Plasma Analysis

Plasma diagnostics with the quadrupole mass spectrometer is performed by sampling plasma particles (ions and neutrals) which stream through a *sampling aperture* located at or near the wall of the plasma reactor. After extraction from the plasma, the flux of particles is mass analyzed. Analysis of the energy of the sampled particles can be added to their mass analysis. A typical experimental set up for plasma analysis with quadrupole mass spectrometry is shown in Fig. 5-4.

The system consists of three main components:

1. The particle *extraction/collimation optics* (ion source, the extraction lens, and as shown in Fig. 5-4, the differential pumping unit described in Section 5.1.2.1)
2. The mass analyzer
3. The ion detector

The most complicated component in the system, in terms of its design and effect on signal collection and interpretation, is the extraction optics.

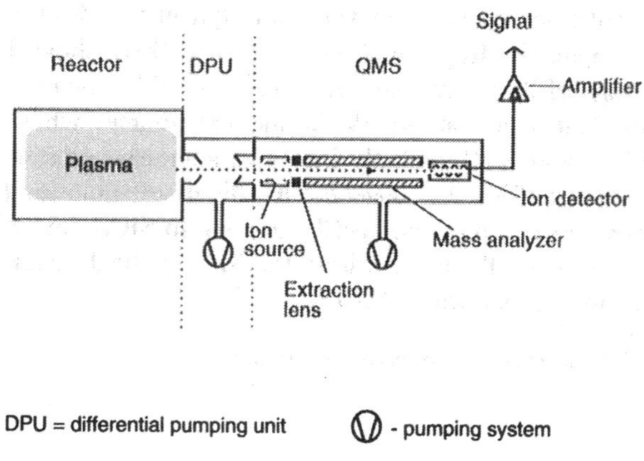

Fig. 5-4 Quadrupole mass spectrometer (QMS) setup for plasma diagnostics.

In sampling neutral species one must cope with the complexity introduced by dissociative ionization that can take place in the ion source. Most molecules undergo dissociation upon ionization, and the same detected ion can be the result of several different precursors. For example, the ion CH_3^+ can result from ionization of neutral species originating from methane, ethane, propane, methyl radicals, and so on.

The ion source of a quadrupole mass spectrometer is usually operated at about 70 eV. An electron energy of such magnitude is sufficient to frequently induce also *fragmentation* of the neutral molecules or radicals entering the source. In plasma diagnostics, the fragmentation in the ion source often results in the formation of fragment ions from one species, which interfere with ions from another, since many of the molecules produced by the plasma are either multimers or very simple fragments of the starting molecules. The analyzed gas may therefore not be identical to the one in the plasma. Thus one has to be careful in the interpretation of results obtained with the mass spectrometer when trying to discern between the radicals originating in the plasma and those that are the result of fragmentation in the mass spectrometer.

To reduce the fragmentation of molecules or radicals extracted from the plasma in the ionization source, it is possible sometimes to operate the source at a reduced voltage. Electron energies of the order of 20–30 eV have been found to be helpful to reduce this effect [15, 16] and simplify the mass spectra in general.

When reducing the energy of the electrons in the ion source, one must choose energies above the ionization potential of the species of interest but below the ionization potential of interfering species. Lowered ionization energies, however, result also in reduced sensitivity. The lower limit of the usable electron energy in the ion source is determined by the requirement to obtain an adequate signal-to-background ratio, since the absolute signal levels are reduced at lower electron energies due to the decrease of the ionization cross-sections.

The extent to which the electron energy should be reduced depends also upon the appearance potential functions of various fragment ions from the molecule of interest. For example, the fragmentation patterns of SiH_4 show that an electron energy in the range of 20 eV reduced the SiH^+ and Si^+ ions to a very low level with respect to their value at 60 eV in the ion source, while the ionization efficiency of SiH_3^+ and SiH_2^+ actually increases somewhat relative to the 60 eV value [17]. A similar effect was observed in silicon tetrachloride ($SiCl_4$) for the ionization efficiencies of $SiCl^+$ and $SiCl_2^+$ relative to $SiCl_3^+$ and $Si\ Cl_4^+$ [16]. In plasmas of Ar and $SiCl_4$, the signals from the $SiCl_x$ radicals saturate at about 30 eV in the ion source, as shown in Fig. 5-5.

5.1.2.1 Flux Analysis of Neutral Species

According to Vasile and Dilla [12], one can distinguish between two types of plasma diagnosis with quadrupole mass spectrometry: *flux analysis* and *partial pressure analysis*. In flux analysis, after sampling the plasma through a small

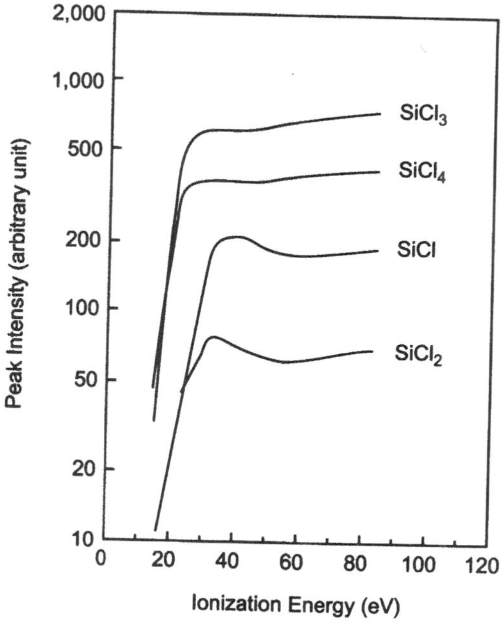

Fig. 5-5 Peak intensities versus ionization energy in ion source in an Ar + $SiCl_4$ plasma (from [16], reprinted with permission from *Plasma Chem. Plasma Process*, vol. 3, p. 235, 1983).

aperture in the plasma reactor, collisionless trajectories lead directly to the ion optics of the mass spectrometer as shown in Fig. 5-4. The analysis is therefore performed on line-of-sight particles forming a molecular beam. This method is best suited for the identification of plasma species and for the optional analysis of their energy.

The size of the sampling orifice should be less than the Debye length so that the presence of the orifice does not disturb the plasma. It should have practically "zero length" to reduce to minimum the effect of collisions with the walls of the orifice.

The flux analysis can, in turn, be differentiated into analysis of flux of neutral particles and flux of ions.

An experimental difficulty that is encountered during flux analysis of plasmas with mass spectroscopy is caused by the difference between the pressures prevailing in the plasma reactor (10^{-3}–10 torr) and those required for the operation of the mass spectrometer (below 10^{-4} torr). To enable the flux analysis of line-of-sight particles, it is necessary to use a pressure reduction stage between the plasma sampling port and the chamber of the quadrupole mass spectrometer. The pressure reduction can be accomplished with a low-conductance connection which is pumped separately as shown in Fig. 5-4 and called a *differential pumping unit, (DPU)*.

A typical differential pumping unit, connected between the plasma and the ion source, has two small low-conductance orifices, several tens of micrometers in diameter, at both ends of the chamber, and is separately pumped by its own vacuum system. The two apertures of the DPU must be aligned with the quadrupole mass analyzer to satisfy the requirement of line-of-sight extraction. The pressure in the differential pumping unit should be reduced below 10^{-4} torr, such that molecules or radicals from the plasma pass through it to the mass spectrometer in a molecular flow regime, with very few collisions. Because the chamber of the spectrometer is pumped by its own vacuum pump, the upper pressure limit can be scaled arbitrarily, and the full dynamic range of the quadupole mass spectrometer is available over the rescaled range.

Flux analysis is the most common technique employed for sampling glow-discharge-type process plasma, since in the majority of applications there are no restrictions on the distance between the plasma vessel and the spectrometer.

5.1.2.2 Partial Pressure Analysis of Neutral Species

In partial pressure analysis, the gases are sampled through a low-conductance connection between the plasma reactor and the mass analysis chamber. In this case, the differential pumping unit is replaced by a leak valve, a sintered porous disk of inert material, or a capillary.

In the arrangement used for partial pressure measurement, collisions take place between the plasma particles and between the particles and the walls of the connecting unit before reaching the mass spectrometer. Condensible or reactive species, especially radicals, react by collisions in the gas phase or on the surfaces or condense on the walls and are lost during the passage to the mass spectrometer. Thus the measured partial pressures may not correspond to those prevailing in the plasma. In addition, the measured partial pressures depend on the pumping speed in the mass spectrometer region.

Partial pressure analysis does not measure line-of-sight plasma particles and hence allows simpler vacuum connections between the mass spectrometer and the plasma chamber and generally requires less stringent differential pumping than flux analysis. The instrumentation for partial pressure analysis can therefore be made simpler than the flux analysis.

5.1.2.3 Ion Analysis

In most plasma diagnostic experiments only neutral species of the plasma are analyzed with the quadrupole mass spectrometer. With suitable ion optics such as special ion lenses, either positive or negative ions can also be extracted from the plasma and analyzed. In such a lens, the filament of the ion source is turned off, and suitable voltages are applied to the ion optics. Since the analyzed ions are produced at the plasma potential, the entire mass analyzer has to be electrically isolated in such a way that it could be adjusted to the potential of various plasmas.

Quadrupole mass spectrometers operate optimally with ions in the energy range of about 5 to 25 eV. If the sampling is done through a grounded wall and the plasma potential is less than 25 V, then the quadrupole mass spectrometer can operate at ground potential. Even in such conditions, ion-neutral collisions and ion formation processes occurring in the plasma sheath can cause the arrival of the ions at the mass spectrometer with a range of energies [18]. The energy distribution of the ions can distort the relative abundance of the observed ions as compared to the true abundance in the plasma.

If the ions are sampled through a negatively biased electrode, or if the plasma potential is large, then it is essential to incorporate energy filtering of the ions before the mass spectrometer. The potential of the mass spectrometer must be at, or slightly below the potential of the sampling aperture to be able to observe ions formed near the orifice.

Flux analysis of ions is not as straightforward as for neutrals, due to the electrostatic interaction between the sampling aperture and the plasma. The flux of the charged particles that passes through the sampling aperture depends on the plasma conditions in the vicinity of the aperture and the aperture's geometry and electrostatic potential. Each specific case has to be addressed individually as there is no common solution to this problem. General conditions for extracting plasma ions from a glow discharge are discussed by Drawin [19] as a function of magnitude of the ion mean free path, λ_i, the Debye shielding length, λ_D, and the diameter of the extraction aperture.

Ion optical lens components (cylindrical lenses in particular) have large apertures and confining surfaces that admit and scatter a solid angle of the neutral flux that is larger than optimal for ideal molecular beam formation [12]. In systems intended for flux analysis of both ionic and neutral species, collimation of incoming neutral flux is difficult to achieve. Ionization of the neutral flux before significant scattering occurs is therefore essential in such systems.

When the experimental arrangements are correctly tuned, the analyzed fluxes of positive ions can be representative of the ion fluxes arriving at the surface exposed to the plasma. However, the measured ion fluxes do not necessarily reflect the fluxes of neutral species in the plasma. Because the large fraction of low-energy electrons in the plasma favors ionization of species with low ionization potentials, relatively higher fluxes of such species will be measured.

5.1.3 Summary

Mass spectroscopy is an excellent technique for detecting neutral particles as well as positive and negative ions from the plasma. Mass spectrometric sampling of a plasma is an intrusive technique because the sampling port has to be in contact with the plasma. The analyzed particle fluxes depend upon the sampling sheath thickness and voltage. Ion-molecule collisions in the sheath region of the sampling orifice and downstream from it may affect the measurement accuracy of densities, especially those of ions.

Disturbances caused to the plasma by the apparatus can be minimized by choosing the right geometry for the orifice and the correct extraction voltage in relation to plasma parameters.

When setting up a quadrupole mass spectrometer for plasma diagnostics, the characteristics of the QMS must be optimized for each of the three possible applications [12]:

1. Flux analysis of neutral particles
2. Flux analysis of charged particles
3. Partial pressure analysis of equilibrated gases

When the sampling aperture and extraction optics are designed correctly, both neutrals and ions representative of the plasma composition near the aperture can be sampled in such a way that the extracted species are not affected by subsequent gas-phase and surface collisions. Plasma diagnostics with mass spectrometry can be used to study the chemical reactions occurring in the plasma and their kinetics and plasma-surface interactions (see, for example, [16, 17, 20–23]). There are examples where the flux analysis of ions and neutrals has provided the most detailed description of material production or modification by plasmas [12].

By changing the position of the sampling aperture, it is possible to perform the plasma diagnostics at different positions in the plasma [24]. The reaction rate constants can be calculated from these measurements, while local rate constants can be derived from the analysis of plasma chemistry as a function of pressure [22].

Mass spectrometry of plasmas can also be used as a tool for end-point detection in plasma etching as, for example, in RIE of silicon nitride, silicon oxide, or titanium tungsten layers over silicon [25]. The detection is based on the occurrence of a very strong change in the signal from specific species at the end-point of the etching, that is, when the surface film has been etched to a depth that exposed the substrate. A 13 times decrease in the signal from CN was observed in Si_3N_4 / Si, a 5 times decrease in the signal from Cl_2 was observed in SiO_2 / Si, and a 23 times increase in the signal from SiCl was observed in TiW / Si etching. The response time for each mass was a few tenths of a second or less, making the technique suitable for controlling the process [25].

The main disadvantages of the quadrupole mass spectrometry analyses are

1. Requirement of careful design of the sampling and extraction optics, which cannot always be made nonperturbing to the plasma
2. Need of differential pumping requirements for maintaining collision-free conditions

The pumping speed of the chamber of the mass spectrometer is also a significant factor to be taken in to consideration since the time required for a complete set of measurements is usually quite long. The pumping speed should be high enough to avoid the buildup of background gas over the duration of the experiment.

5.2 ELECTROSTATIC PROBES

Electrostatic or *Langmuir probes* are another tool for plasma diagnostics suitable for measuring parameters of cold plasmas. This technique allows the determination of plasma density, n, electron temperature, T_e, plasma potential, V_p, and floating potential, V_{fl}. The probes do not provide information on the chemistry of the plasma, but serve well in complementing the mass spectroscopy diagnostics.

A Langmuir probe is a metallic insulated electrode, except at the tip, which is exposed to the plasma. The probe is made of a high-melting-point metal such as tungsten, molybdenum, or platinum. Tungsten wire is commonly used. The probe is inserted into the plasma reactor through an electrically insulated seal. The insulator of the probe should be chemically inert at high temperatures. It can be either quartz or a vacuum-compatible ceramic, such as high-purity alumina. The tip of a cylindrical probe exposed to the plasma, as illustrated in Fig. 5-6, is several millimeters long and less than 1 mm in diameter. The vacuum seal can be fixed or can permit changes in the position of the probe.

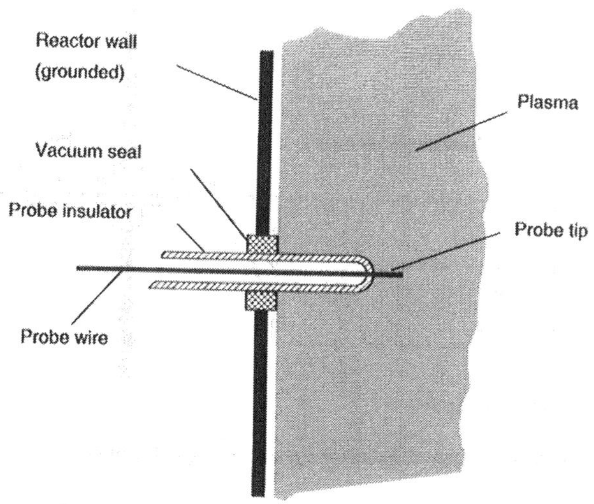

Fig. 5-6 Diagram of single Langmuir probe.

Details on the use of electrostatic probes for plasma diagnostics can be found in monographs dealing with plasma diagnostics in general and probes in particular [4, 26–29]. Due to the fact that the measurements are made by inserting a probe in the plasma, this technique is an in-situ intrusive diagnostic method that locally perturbs the plasma.

An electrode inserted in the plasma changes the physical conditions in its vicinity, which means that the measurement is done in a plasma disturbed by the probe itself. The probe affects the plasma mainly by changing the electric field, thereby the particle distribution and energy. The magnitude of this disturbance depends on the dimensions of the probe and the properties of the plasma. Probes that satisfy the relation $r > \lambda_D$, where r is the radius of the probe, are called *thin sheath probes*. These are conventional type probes used in diagnostics of cold plasmas.

Most used are *single* and *double Langmuir probes*. These are operated by applying an external variable potential, V, and by measuring the current through the probe as a function of the external potential; the *I-V characteristic* of the probe is thus obtained. The applied potential is usually swept along a sawtooth-shaped time dependence. For a single Langmuir probe, the voltage V is referenced to a large metal surface in the plasma chamber or to the walls of the chamber itself. For the double Langmuir probe, also called *double floating electrostatic probe*, (*DFEP*), the voltage is applied between the two electrodes, both insulated from the reactor.

The experimental arrangement is simple, requiring in addition to the probe, a scanning DC power supply (usually $\pm$ 100 V) and an X-Y recorder. A typical circuit diagram for the probe is shown in Fig. 5-7.

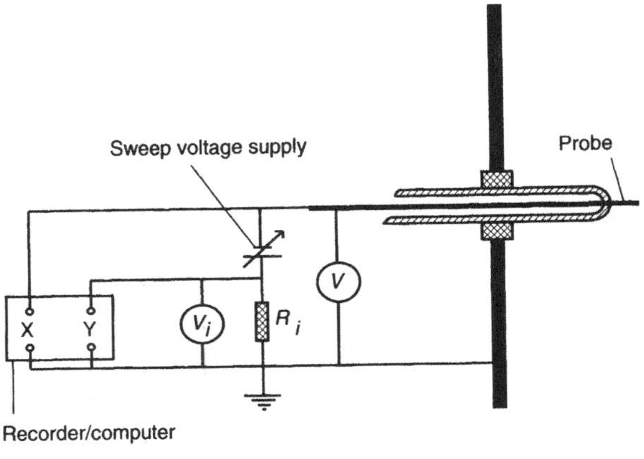

Fig. 5-7 Typical electrical circuit for single Langmuir probe.

The X-Y recorder traces the I-V characteristic of the probe similar to the one illustrated in Fig. 5-8. The voltage on the probe is recorded on the x-axis, while the current through the probe is recorded on the y-axis, as the voltage drop across a reference resistor R_i. The X-Y recorder can obviously be replaced by a microcomputer to log the data and perform its analysis.

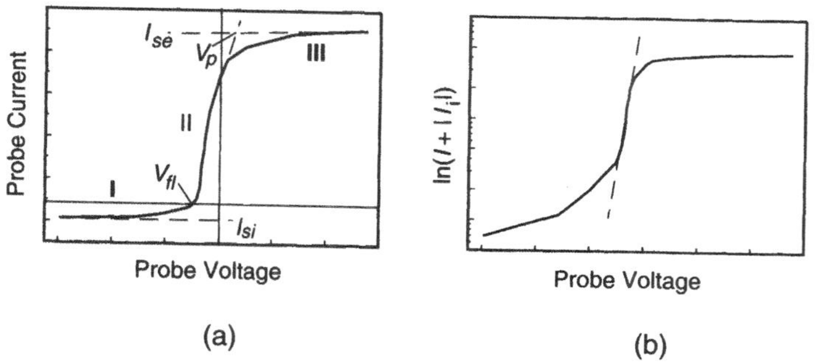

Fig. 5-8 I-V characteristic of a single Langmuir probe: (a) linear display; (b) semilogarithmic display.

5.2.1 Probe Characteristics

Consider a plasma containing only two species: electrons and ions of equal and opposite charge. In an unperturbed plasma, according to Eqs. (1.32) and (1.34) of Sec. 1.3, the mean ion speed, $\bar{v}_i$, is much smaller than the mean electron speed, $\bar{v}_e$, by the ratio $(m_i/m_e)^{1/2}$. Therefore, the flow of electrons to the tip of the probe is higher than the flow of ions and the total current from a probe of area A_p inserted in the plasma will be [5]

$$I = -eA_p\left(\frac{1}{4}n_i\bar{v}_i - \frac{1}{4}n_e\bar{v}_e\right) \approx \frac{1}{4}eA_pn_e\bar{v}_e > 0 \tag{5.2}$$

The probe current is dominated by the electrons, and the probe emits a net positive current. If the probe is electrically insulated from the other parts of the plasma reactor, that is, if the probe is a floating one, it will rapidly charge up negatively, until the electrons are repelled and the net current brought to zero. The tip assumes a negative potential called the *floating potential*, V_{fl}. A probe connected to a high-impedance oscilloscope will measure the floating potential.

The *plasma potential*, V_{pl}, is defined as the potential at which all electrons arriving near the probe are collected and the probe current equals the electron current. If the probe is at the plasma potential, the perturbation it induces to the free ion and electron currents will be small. The plasma potential, V_{pl}, is in fact the electric potential in the plasma in the absence of a probe.

At potentials higher than V_{pl}, the electron current cannot increase any further and the *electron saturation current*, I_{se}, is measured; see Fig. 5-8(a). When the probe potential decreases to values below V_{pl}, it becomes negative relative to the plasma, and an increasing number of electrons are repelled from the probe. At sufficiently low potentials, the electron and ion current densities to the probe are equal, $J_e = J_i$, and the total probe current is zero. The probe potential at this point is, as mentioned before, the floating potential, V_{fl}. At potentials lower than V_{fl}, only ions are collected from the plasma, and the *ion saturation current*, I_{si}, is measured; see Fig. 5-8(a).

As a result of the behavior just described, the I-V characteristic of a single Langmuir probe has the general shape shown in Fig. 5-8. Corresponding to this description, three regions can be observed in the probe characteristic: the ion saturation region I, the transition region II, and the electron saturation region III.

The loads of the measuring devices distort the described ideal I-V characteristic of the probe. The floating potential can be measured directly only by a probe with an infinite impedance, while the saturation current can be measured only with a probe of zero impedance. The floating potential of the plasma, V_{fl}, is defined as corresponding to the point in Fig. 5.8 (a), where the current through the probe is zero. Its value is in fact identical to the sheath potential defined in Eqs. (1.28a) and (1.28b) in Sec. 1.3.4, which for a spherical surface is given by [27]

$$V_{fl} = -\frac{kT_e}{2e}\ln\left(\frac{\pi m_e}{2m_i}\right) \tag{5.3}$$

The plasma potential, V_{pl}, is determined from the intersection between the extrapolations of the sections of the curve in the transition region II and the electron saturation region III. Its value can be estimated from

$$V_p = kT_i\,\frac{\ln 2}{q_i} \tag{5.4}$$

where T_i = temperature of ions involved
q_i = charge of ions

Consistent with the higher mobility of electrons, the electron saturation current is about two orders of magnitude higher than the ion saturation current.

5.2.1.1 Effect of Magnetic Field

So far, the description of the probe behavior has assumed that no magnetic field is present and that the motion of the charged particles in the plasma is controlled only by the electric field. If, however, a magnetic field is present in the plasma, the electrons and ions will no longer move in straight lines, following the electric field lines, but will orbit instead around the magnetic field lines, in circular orbits with a radius equal to the *Larmor radius,* defined by Eq. (2.13) in Sec. 2.4:

$$r_L = \frac{mv_\perp}{eB} \tag{5.5}$$

As a result, the motion of the particles across the magnetic field is strongly restricted, while the motion along the magnetic field line is almost unaffected. The effect of the magnetic field can be neglected only if $r_L \gg a$, where a is the typical dimension of the probe ($a = r$ for cylindrical probes).

According to Eq. (5.5), the Larmor radius of the electron, r_{Le}, is smaller than that of the ion, r_{Li}, by the factor m_e/m_i; hence the electrons are much more affected than the ions by the presence of the magnetic field. Because normally $r_{Li} > a$, the ion saturation current is almost not affected by the magnetic field. On the other hand, the electron saturation current measured by a circular probe, which is parallel to the magnetic field, can be strongly reduced, as illustrated in Fig. 5-9. The electron saturation current measured with a probe positioned normal to the magnetic field is much less affected (Fig. 5-9).

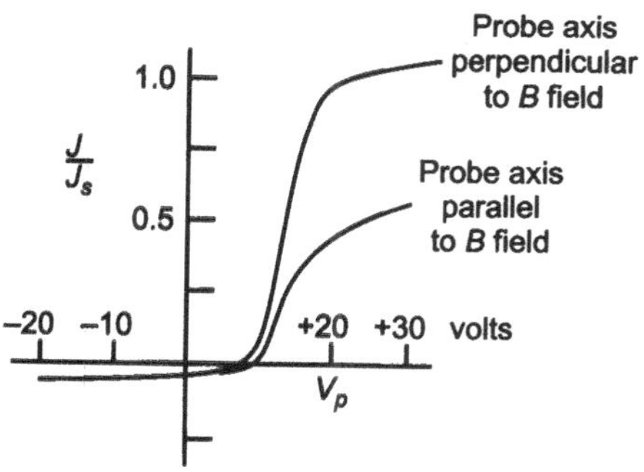

Fig. 5-9 Effect of magnetic field on probe characteristic (from [4], reprinted by permission of John Wiley & Sons, Inc.).

5.2.2 Probe Analysis

Although the experimental arrangement for probe measurements is relatively simple, one has to be careful when analyzing the experimental data.

The only probe regime that can be reasonably understood is the pressure region $p \leq 0.1$ torr, in which the mean free path λ is much greater than the probe size, which in turn is much greater than the Debye length, $\lambda \gg r > \lambda_D$ [30]. However, for phenomenological estimations, or for the determination of the functional relationship between the parameters of the plasma and the discharge conditions (i.e., power, pressure etc.), the Langmuir probe is a relatively simple, valuable diagnostic tool.

An exact interpretation of the results of the probe measurements for absolute determination of the plasma parameters may be more complicated than described here. The following presents only a simplified description of the evaluation of the measurements by single and double Langmuir probe diagnostics.

5.2.2.1 Single Langmuir Probes

As illustrated in Fig. 5-8, increasing the probe voltage, V, beyond V_{fl} results in a steep rise in the electron current in region II of the I-V characteristic. In this region, the electron current of the probe follows the relation [4]

$$\ln I_e = \frac{V}{kT_e} + \ln A_s J_o \tag{5.6}$$

where A_s = the area of the probe sheath
J_o = the random electron current density
and kT_e is expressed in electron volts and V in volts.

For a typical cold plasma with $T_e = 1$ eV and $n_e = 10^{11}$ cm^{-3}, the Debye length is $\lambda_D = 20$ μm, so that the Debye length is usually much smaller than typical probe dimension of millimeters ($\lambda_D \ll a$). As the thickness of the plasma sheath is commonly much smaller than the radius of the probe, one can assume that $A_s = A_p$, A_p being the area of the probe.

The total probe current, I, is the difference between the electron and ion current. Therefore,

$$I_e = I + |I_i| \tag{5.7}$$

Equation (5.6) can therefore be changed to

$$kT_e = \frac{dV}{d[\ln(I + |I_i|)]} \approx \frac{\Delta V}{\Delta[\ln(I + |I_i|)]} \tag{5.8}$$

and the electron temperature, T_e, can be determined from the slope of the logarithm of I_e versus probe voltage in region II (Fig. 5-8 (b)).

Increasing the probe voltage beyond V_{pl}, the electron current reaches the saturation value I_{se}. The saturation current, according to Eq. (5.2) is given by [4]

$$I_{se} = \frac{1}{4} eA_p n_e \left(\frac{2kT_e}{m_e}\right)^{1/2} \tag{5.9}$$

and with T_e determined from Eq. (5.8), the electron density n_e can be calculated from I_{se}.

For $T_i \ll T_e$, which is characteristic of cold plasmas, the saturation ion current, I_{si}, is nearly independent of the ion temperature and is given by [4]

$$I_{si} = 0.4n_i eA_p \left(\frac{2kT_e}{m_i}\right)^{1/2} \tag{5.10}$$

With T_e established from Eq. (5.8), the ion density n_i can be evaluated from Eq. (5.10).

The calculation of particle densities from Eq. (5.9) and Eq. (5.10) assumes that the area through which the probe collects the ions is known. However, the ion collecting area may be sometimes significantly different from the geometrical area of the probe, and the estimate of the ion density may be inaccurate. The problem is aggravated in the presence of a magnetic field, especially for the ions.

One has to remember that secondary electrons emitted from the probe by the collected ions also contribute a positive current that is indistinguishable from the ion current. This addition to the current is another source of error in the calculation of n_i.

5.2.2.2 Double Floating Probes

Double probes are used for diagnostics of RF and MW discharges when a reference electrode is not available, as in the case of inductively coupled RF systems. The double probe technique is also useful in a magnetized plasma. A double probe is composed of two, usually identically tipped, probes separated by a fixed distance, as illustrated in Fig. 5-10. Figure 5-10 also shows the electrical circuit, the momentary potential distribution between probes and an ideal I-V characteristic for double probes of identical areas.

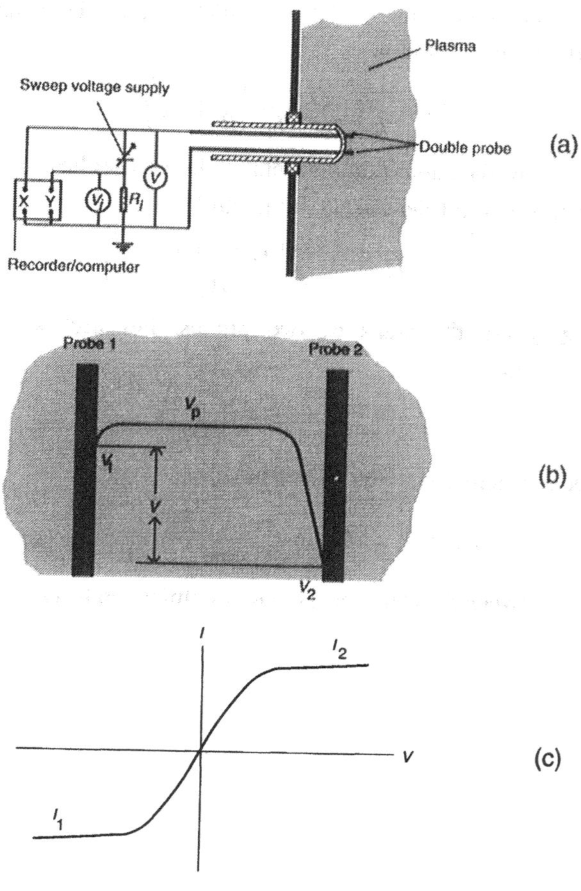

Fig. 5-10 Double Langmuir probe: (a) probe and electrical circuit; (b) momentary potential distribution between probes; (c) I-V characteristic of identical probes.

The tips of the probes have to be far enough from each other so that the plasma sheaths of the probes would not overlap, but close enough to sample the same region of the plasma. Both probes are insulated from the ground, float with the plasma, and are unaffected by changes in V_{pl}. The circuit is equilibrated when the electron current flowing to one probe is balanced by the ion current flowing to the other. The currents are therefore limited by the ion saturation currents for each probe, and because the ion current is much less affected by a magnetic field, the double probes are preferred when magnetic fields are present.

In an experimental arrangement, a sawtoothed time variable voltage is applied to the probe, as shown in Fig. 5-10. From the I-V characteristic, plasma parameters can be calculated as follows [23, 27]:

- The electron temperature can be calculated from the slope of the I-V characteristic at the origin,

$$T_e = \frac{e}{k}\left(\frac{I_1 I_2}{I_1 + I_2}\frac{dV}{dI}\Big|_{I=0}\right) \tag{5.11}$$

where $I_{1,2}$ are the saturation currents of the two probes.

- The electron current density, J_e, is given by

$$J_e = -\frac{1}{2}\left(\frac{I_1 A_2 + I_2 A_1}{A_1 A_2}\right) \tag{5.12}$$

where $A_{1,2}$ are the areas of two probes. For probes of equal areas $A_1 = A_2 = A_p$,

$$J_e = -\frac{1}{2}\left(\frac{I_1}{A_p}\right) \tag{5.13}$$

- The electron density is then calculated from

$$n_e = -1.22\frac{J_e}{e(8kT_e/\pi m_e)^{1/2}(\mu_e/\mu_i)^{1.08}} \tag{5.14}$$

The ratio between the electron and ion mobilities can be estimated from [27]

$$\mu_e/\mu_i = 7.64\, m_i^{0.46} \tag{5.15}$$

where m_i is the mass of the ion.

The choice of m_i is straightforward for a plasma containing only one type of ions; otherwise, the mass of the most abundant ion can be used for the calculation.

5.2.2.3 General Remarks

In the discussion of single or double probes analysis, the possible existence of negative ions in the plasma has been neglected. Plasmas of halogen-bearing gases or oxygen have significant concentrations of negative ions, and in such cases the contribution of the negative ions has to be considered in the analysis.

When using Langmuir probes to evaluate the plasma parameters, certain experimental considerations should be kept in mind:

1. Insulating layers may form on the probe in a plasma used for deposition of insulators or due to contamination.
2. The probe may become short circuited or its effective area may increase if a conductive material coats the insulation of the probe in a plasma used for deposition of conductive films.
3. The collected data may be influenced by thermal effects.

The distortion of probe characteristics by contaminated probes is particularly strong in the region close to the plasma potential. The ion current at highly negative probe potentials is less sensitive to probe contamination. Some of these problems may be reduced by cleaning the surface of the probe. This can be done through ion bombardment, by biasing the probe to a relatively large negative potential of about 100 V. Another way used for cleaning the probe is by passing a high-electron current through it which heats it to a high temperature.

The voltage sweep on the probe has to be fast because rapid changes may occur in the plasma during the sweep. But if the sweep rate is too fast, the probe sheath may not have enough time to come to equilibrium and displacement currents may occur due to the capacitance of the probe and the connecting leads. Sweep rates of 1 to 10 V/μsec are commonly used in probe diagnostics.

Electrostatic probes are a convenient and often-used tool for plasma diagnostics, for determination of electron temperature and plasma density. A detailed description of the different types of electrostatic probes and their analysis can be found in a recent review by Hershkowitz [29].

5.3 OPTICAL METHODS

Optical diagnostic techniques, which use photons to transmit the information from the plasma to the detection tool, are the least intrusive in-situ plasma diagnostic methods. The photons in most cases cause only a negligible perturbation to the process. The photon-based diagnostics include the often-used emission spectroscopy, absorption spectroscopy, and laser induced fluorescence, but also Raman scattering, coherent anti-Stokes Raman spectroscopy (CARS), optogalvanic effects, laser interferometry, and ellipsometry.

The optical diagnostic techniques can provide information about the concentration of the different species and their temperature without perturbing the plasma. They are based on the analysis of optical radiation in plasma, which is mostly a result of the capture of free electrons into various excited states of the atoms and ions and from acceleration of electrons during collisions with ions and atoms (*bremsstrahlung*).

The different types of optical spectra observed in cold plasmas and their sources are summarized in Table 5-2. It can be seen that the spectra can contain

TABLE 5-2 Spectra Observed in a Plasma (from [31], reprinted by permission of John Wiley & Sons, Inc.)

Particle	Degree of Freedom	Type of Spectrum	Spectral Region
Atom or ion	Electronic excitation	Line	UV-visible-IR
	Ionization	Continuum	UV-visible-IR
	Translation	Line profiles	
Electrons	Recombination	Continuum	UV-visible
	Free-free transitions	Continuum	IR
Molecules	Rotation	Line	Far infrared
	Vibration-rotation	Band	IR
	Electronic excitation	Band systems	UV-visible-IR

individual spectral lines, band, or continuum radiation and span the range from the far infrared (IR) to ultraviolet (UV). Soft X rays are also existent to some extent in cold plasmas characterized by high T_e. Radiations in the visible and IR spectra can be used to identify excited plasma species. Absorption measurements with a suitable radiation source can be used to evaluate the population of non-radiating metastable levels of first excited states that radiate in vacuum UV.

Laser interferometry does not provide chemical information, but it does give an extremely accurate measurement of material deposition or etching rates. In-situ ellipsometry can be used for measuring small thickness changes or to characterize very thin layers (< 30 Å).

Several optical diagnostic methods are discussed in more detail next.

5.3.1 Optical Emission Spectrometry

Optical emission spectrometry (*OES*), which is the spectral analysis of the light emanating from a plasma, is probably the most widely used method for monitoring and diagnosis of plasma processes. By measuring the wavelengths and intensities of the emitted spectral lines, one can identify the neutral particles and ions present in the plasma. The method is implemented both in research laboratories and in manufacturing for production control.

The spectral fingerprint of optical plasma emission provides information about the chemical and physical processes that occur in the plasma. This technique has the advantage of being external to the reactor and vacuum system and provides besides spatial and temporal resolution also high reliability. However, the OES technique is limited to the monitoring of light-emitting species, and the emission intensity is not always directly related to the concentration of the species in the plasma.

Optical emission can be produced in a plasma as a result of electron impact excitation or dissociation

$$A + e \rightarrow A^* + e \tag{5.16}$$

$$AB + e \rightarrow A^* + B + e \tag{5.17}$$

or by an ion impact process,

$$A^+ + e(+M) \rightarrow A^*(+M) \tag{5.18}$$

which create an excited species A^*, followed by the release of energy of the excited species A^*:

$$A^* \rightarrow A^{**} + h\nu \tag{5.19}$$

In the foregoing equations, the sign * indicates the excited, emitting species; $e(+M)$ designates a neutral species, a negative ion, an electron plus a third body, or a surface in contact with the plasma [32]; and A^{**} is either the ground state of the atom or another excited state at an energy level lower than A^*. For example, the reaction described by Eq. (5.16) is responsible for radiation emission from excited fluorine atoms in $CF_4 + O_2$ discharges or from Cl in Cl_2 discharges, while the process described by Eq. (5.17) has been observed by time-resolved emission in the momentary cathode sheath of a radio frequency Cl_2 discharge [32]. Excitation of hydrogen atoms in a DC discharge of hydrogen has been ascribed in part to the reaction described by Eq. (5.18).

Molecular species can also be excited by chemiluminescent recombination reactions:

$$A + BC \rightarrow AB^* + C \rightarrow AB + C + h\nu \tag{5.20}$$

Such processes occur in Cl_2 discharges due to recombination of Cl atoms. In discharges containing fluorine atoms in the presence of silicon, optical emission results from the reaction

$$SiF_2 + F\,(\text{or } F_2) \rightarrow SiF_3^* + (F) \tag{5.21}$$

Metastable species can also excite emission in plasmas as is the case, for example, in the enhancement of H-Balmer-α emission caused by addition of Ar to H_2 discharges observed by Krogh et al [33].

Usually, the most intense radiation emitted from the plasma originates in the transition from the first excited state, W_1, to the ground state, W_0, of the particle. As every particle (atom, molecule, ion, radical) has precisely defined energy levels, each emits a characteristic spectral line of frequency:

$$\nu_{10} = \frac{W_1 - W_0}{h} \tag{5.22}$$

and wavelength

$$\lambda_{10} = \frac{hc}{W_1 - W_0} \tag{5.23}$$

where c is the velocity of light.

The apparatus required for OES is one of the least complicated. It necessitates a monochromator to disperse the light emitted from the plasma, a set of optics to image the light into the detector, and a photodetector to measure the dispersed light. A typical experimental arrangement is presented in Fig. 5-11.

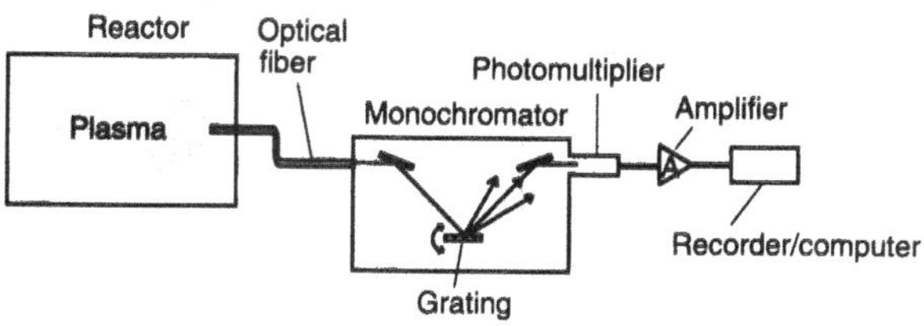

Fig. 5-11 Experimental setup for optical emission spectroscopy of plasmas.

The light emitted by the plasma is conveyed to the monochromator either directly or via a fiber optic cable. Diffraction at the grating of the monochromator disperses the light into its spectral components. Only the light of a selected wavelength passes through the exit slit of the monochromator and strikes a photomultiplier detector where it produces a photocurrent proportional to the intensity of the incident light. The spectrum of the plasma emission is obtained by rotating the grating of the monochromator with a motor.

Monochromatic light emerging from the exit slit of the monochromator can be detected with a photomultiplier tube used in analog mode (current measuring) or digital mode (photon counting). As an alternative, the exit slit can be replaced with an *optical multichannel analyzer* (OMA) placed in the plane of the exit slit. The advantage of an OMA is that a portion of the emission spectrum can be recorded simultaneously without scanning the monochromator, thus providing much faster data collection. Although the emissions from the plasma span a very large range of frequencies, from the infrared to the soft X ray, the spectra most widely used in OES span the range between 200 and 900 nm.

Identification of ions in the plasma by optical emission spectroscopy requires a detector able to monitor lines with wavelength $\lambda < 200$ nm, because ion transitions occur at these shorter wavelengths. For ion identification, the monochromator must be maintained under vacuum to avoid absorption of radiation at these short wavelengths. Fiber optics are not available at these wavelengths. For observation of lines with $\lambda < 110$ μm, the monochromator must be connected directly to the plasma chamber through a hole, because there are no materials with sufficient optical transmission at these wavelengths from which to fabricate the lenses. Even mirrors cannot be used due to reflectances lower than 10% at these wavelengths. Instead, a concave grating, which performs both the task of diffraction and optical imaging, is used.

A spectrometer resolution of 0.1 to 1 nm is required for identification of species. A monochromator of 0.25-m focal length can provide ~ 0.05-nm resolution, sufficient for most studies. When higher resolution is required to detect atomic fine structure or molecular rotational levels, a monochromator of long focal length (e.g., 1-m focal length with 0.01-nm resolution) or a Fabre-Perot interferometer (~ 0.0005-nm resolution) can be used.

The species emitting a certain radiation can be identified from the specific frequency of the radiation, using tables of emission spectra. However, the use of OES for plasma diagnostics is complicated by several factors. One of them is the extreme complexity of the spectra. Emission from atomic and diatomic species provides the most useful information in OES diagnostics because they have simpler narrower spectra than the polyatomic species. The last ones have very broad spectra due to their vibrational and rotational transitions or do not radiate significantly in the spectral range used. While identification of elemental lines is relatively easy, the identification of lines of chemical compounds is difficult due to the large number of emission lines produced by each element in the compound. Identification of weaker lines, or identification of plasma species of low density, is complicated because the main components of the plasma possess many spectral transitions. The optical signals from the main components may thus obscure those from less abundant species.

Another difficulty in the use of OES in plasma diagnostics for quantitative analysis is the dependence of the optical emission intensity on the ability of the plasma to excite the ground state species into electronically excited emitting species A^*.

For quantitative evaluation of the emission spectra a first assumption will be that the *emission intensity*, In_s, from a specific species is proportional to its concentration, [F]:

$$In_s = a_s^e [\mathrm{F}] \tag{5.24}$$

where the proportionality factor a_s^e is constant. The proportionality constant a_s^e is given by [34]

$$a_s^e = K \int_0^{\infty} Q(p, n_e) \sigma_s^e(\varepsilon) N_e(\varepsilon) d\varepsilon \tag{5.25}$$

where K = a constant depending on the sensitivity of the detector
$\sigma_s^e(\varepsilon)$ = cross section for excitation of the emitting species to a given excited state caused by the impact of an electron of energy ε
$N_e(\varepsilon)$ = number of electrons in the energy range $d\varepsilon$ present in the volume of the reactor viewed by the detector
$Q(p, n_e)$ = quantum yield for emission from the given excited state as a function of discharge pressure and electron density

In principle, Q varies between 0 and 1.

According to Eq. (5.24) and (5.25) the factor a_s^e is obviously not a constant. Because $N_e(\varepsilon)$ depends on the applied RF power and gas mixture and Q depends on the total pressure and gas mixture, a_s^e will never be exactly a constant.

If no significant changes are made in the plasma parameters, it can be still assumed that the excitation efficiency of the plasma is constant. If any modifications are made in the plasma parameters such as power, pressure, or gas composition, this assumption is no longer valid, and the intensity of the emitted light cannot be assumed to be proportional to the density of ground state species.

Furthermore, the foregoing equations assume that all the emissions stem from excitation of the ground state of the species. Whenever other processes contribute to In_s, then Eq. (5.24) is completely invalid. Measurements by Mogab et al. [35] showed that a_F^e for fluorine atoms can vary by over a factor of two when oxygen is added to a CF_4 plasma.

It is therefore not possible to draw any direct conclusion about the concentration of the species from the emission intensity In_s. In practice, one must empirically determine a significant spectral line or combination of lines from which the quantity of interest can be derived. This approach is used for end-point detection in plasma etching. At the end point, when the upper film is etched through, pronounced changes occur in the emission lines of the plasma. These changes serve as a stop signal for the etching process. Optical detection of In emission provided, for example, excellent end-point detection during etching of /In/GaAsInP/InP with CH_4-H_2 plasma [36].

However, in some cases of practical interest, the situation may be somewhat simpler than indicated above. In plasma etching applications, the pressure and electron density are relatively low; thus $Q(p, N_e)$ may be close to unity over a range of operating conditions, especially for short-lived excited states [34]. Similarly $\sigma_s^e(\varepsilon)$ may be a relatively "slow" function of electrons energy, ε, and $N_e(\varepsilon)$ may not depend strongly on gas composition. Under such conditions a_s^e may be almost constant. A narrow Doppler width (Fig. 5-12(a)), which corresponds to the measured gas temperature, is evidence for emission caused by the reaction described by Eq. (5.16). A broadened emission (Fig. 5-12(b)) corresponds to other pathways similar to those described by Eqs. (5.17)–(5.20).

Optical emission can also provide information on electron energy distributions in RF discharges. More detailed description of OES and other optical plasma diagnostic methods is given by Donnelly [32].

5.3.1.1 Actinometry

To overcome the problem of nonconstancy of the proportionality coefficient a_s^e in Eq. (5.24), actinometry diagnostics has been developed by Coburn and Chen [37]. This is a relatively simple procedure designed to remove the effect of plasma excitation efficiency from the OES data. A recent review of actinometry and its application for plasma polymerization studies was published by d'Agostino and Illuzi [38]. In this technique, a small quantity (1–2%) of an inert gas, which serves as *actinometer*, is mixed into the gas fed to the plasma. It is assumed that the actinometer does not change the characteristics of the plasma to which it is added. The concentration of the actinometer gas is kept constant and the optical emission derived from it is used as a measure of the excitation efficiency of the plasma when the plasma parameters are changed.

In actinometry, one monitors in the same plasma both the emission In_r from the reactive species and the emission In_i of the inert actinometer gas. If the quantum yield and the cross sections of the reactive and inert species have the same energy dependence, the following relation exists:

$$\frac{In_r}{In_i} = C\,\frac{[F]_r}{[F]_i} \tag{5.26}$$

where C is a constant not affected by plasma parameters.

Equation (5.26) assumes that the cross sections for excitation of the actinometer and the active species are equal. Although there might be differences between the values of the cross sections, the error introduced is not large [38]. A larger error can be introduced if the excitation threshold of the actinometer is significantly different from that of the active species. Actinometry can therefore be used only for emissions from species that have excitation thresholds close to that of the actinometer, a condition, for example, satisfied for Ar and F [37]. The excited state of argon lies at $\sim$ 13.5 eV compared to $\sim$ 14.5 eV for the corresponding fluorine state, and the excitation efficiency is approximately equal for the two atoms. Different actinometers have to be used to study different active species. Coburn and Chen [37] first studied F atoms with Ar as actinometer, while d'Agostino et al. [39] used actinometry for studies of F and O atoms and CO and CO_2 molecules using Ar and N_2 as actinometers.

Although actinometry is not an accurate technique, it can provide a good approximation for the evaluation of densities of species. Actinometry does not, however, solve the problem associated with emissions originating from dissociative processes. The spatially resolved quantity In_{Ar} can provide information about the uniformity of plasma density. In an RF plasma, concentrations and energies of electrons and ions change dramatically during an RF cycle, leading to a high degree of modulation in emissions excited by ion and electron impact. Ion modulation becomes limited at ≈ 3 MHz for typical ions (Cl_2^+, CF_3^+, etc.), while electron energy can be modulated up to ≈ 100 MHz. As a result, time-resolved emission spectroscopy can distinguish between emissions caused by electron impact (Eq. (5.16)) and those caused by impact of heavy particles such as described by Eq. (5.20).

The process described by Eq. (5.16) can also be distinguished from other emission processes by measurement of line shapes.

The light-emitting particles possess translational velocities that cause a Doppler shift to the frequency of the radiation. Since the electron mass is very small compared to atomic or molecular masses, the velocity distribution of the emitting species formed by the reaction described by Eq. (5.16) is the same as for species in the ground state. The velocity distribution of the particles in the ground state is, in turn, determined by the gas temperature. Excited species formed by other impact processes will be created with significantly different translational energy as illustrated schematically in Fig. 5-12.

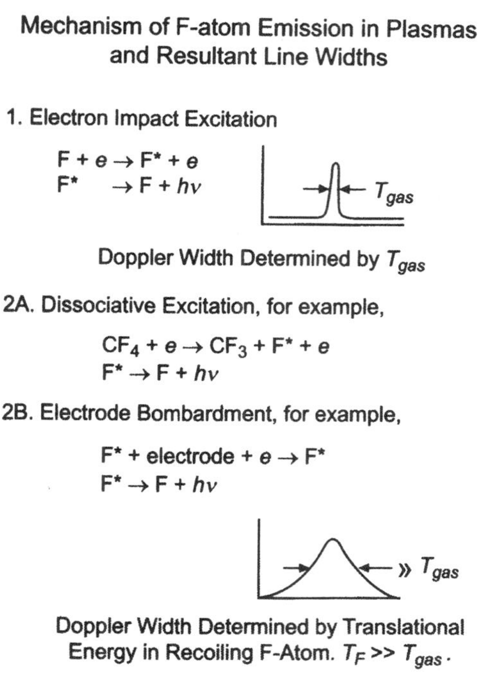

Fig. 5-12 Effect of collision type on Doppler widening of emitted lines (from [32], reprinted with permission from *Plasma Diagnostics*, vol. 1, 1989).

5.3.2 Absorption Spectroscopy

The major drawback of the emission spectroscopy is that it can only provide information about species that are already excited and that may form only a small fraction of the total number of species in question. Absorption spectroscopy, on the contrary, provides for this deficiency and can supplement the optical emission diagnostics. Absorption spectroscopy, however, requires higher resolution because otherwise sharp absorption lines may be lost in the background. Cold plasmas are quite amenable to absorption spectroscopy diagnostics, yet relatively few plasma diagnostic studies have been performed with this technique.

The main advantage of the absorption spectroscopy technique is its ease of implementation in most cases. If light is transmitted through a plasma, its intensity, *In*, at a given wavelength, is given by Beer's law,

$$In = In_o e^{-\sigma n l} \tag{5.27}$$

where In_o = intensity of incident light at the considered wavelength
σ = cross section for absorption
n = density of absorbing species
l = optical path length

If the absorption cross sections of the different species are known, the absolute concentrations can be calculated from the intensity of the transmitted light.

The experimental setup used for optical emission measurements can easily be adapted for absorption measurements by adding an external light source. Light from the source, which can be a lamp or a laser, is directed through the discharge onto the entrance slit of the monochromator. The optimum light source depends on the species that have to be detected. For diatomic and polyatomic species, a continuum light source is best suited. A deuterium lamp can be used in the ultraviolet range of 160–350 nm, while a tungsten lamp is suitable in the visible region. The output of the lamp can be chopped and combined with phase-sensitive detection to suppress the large optical background from the plasma.

The disadvantages of absorption spectroscopy are that it requires vacuum ultraviolet sources for many elements of interest, it has a low sensitivity in the infrared region of the spectrum, and spectrums of multicomponent mixtures can overlap.

5.3.3 Laser Induced Fluorescence

Particles excited by absorption of radiation may undergo radiative transition in which light of the same or different frequency is emitted. When the transition occurs between states of the same multiplicity, the particle is said to emit fluorescence radiation. The intensity of the fluorescent radiation is proportional to the particle density. In principle, the density of molecules, atoms, and ions in the ground state, as well as in metastable or unstable, excited states, can be determined by measuring the intensity of the fluorescence. Due to the fact that fluorescence times are much shorter than a microsecond, it is necessary to use laser pulses with a width of the order of nanoseconds for the excitation of the fluorescence. This diagnostic technique is called *laser induced fluorescence* (*LIF*). Laser induced fluorescence is the most used diagnostic tool of the laser-based techniques; further discussions on LIF can be found in review papers such as [32, 34, 40].

The LIF technique employs a laser whose frequency is tuned to excite a resonance between a lower electronic state and an excited electronic state of a chosen species. A tunable dye laser is generally used to tune the frequency of the exciting light. Continuous tunable dye lasers are pumped by N_2 lasers, Nd-YAG lasers, excimer lasers, or flash lamps for pulsed output or by Ar^+ and Kr^+ lasers for CW radiation. Dye lasers, pumped by either an excimer laser or a Nd-YAG laser are highest in peak power and are the most common tunable pulsed sources. Excimer laser pumped dye lasers provide output from 330–900 nm, while frequency doubled and tripled Nd-YAG lasers can be used to pump dye lasers between ~ 380 and 900 nm.

The experimental setup is illustrated in Fig. 5-13. The laser is usually placed in a horizontal position, parallel to the surface of the discharge electrode. By

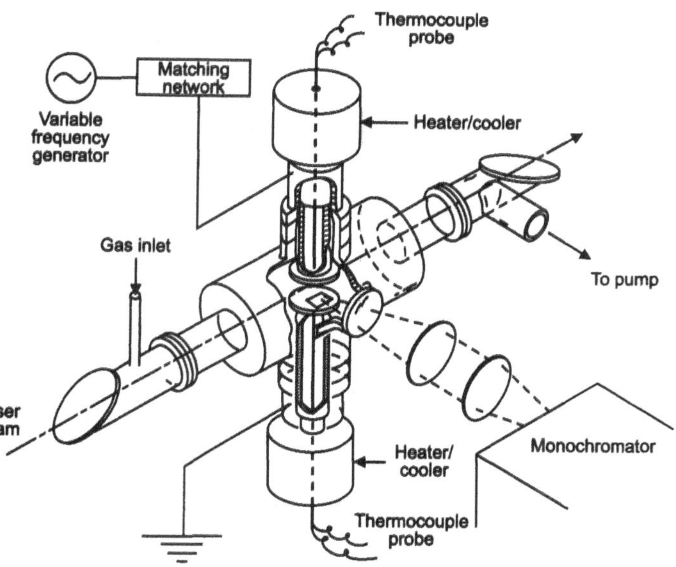

Fig. 5-13 Experimental setup for laser induced fluorescence in a parallel plate plasma reactor (from [32], reprinted with permission from *Plasma Diagnostics*, vol. 1, 1989).

such an arrangement the discharge can be probed in a direction perpendicular to the electrode surface by translating the laser beam and detection optics. The fluorescence is detected at right angles with a monochromator or a filter in front of a photomultiplier tube or by an optical multichannel analyzer. To discern between the fluorescence emission and the plasma background, the spectrally resolved emission is detected and amplified using a boxcar integrator with gated detection. In this arrangement, the signal is recorded only during a very narrow time gate, which is selected to coincide with the timing of the laser pulse, thus identifying only the LIF emissions.

To use the LIF technique, the investigated species must fluorescence with a reasonable quantum efficiency, and the tunable laser must be able to match a transition for the investigated species. The last requirement is impeding the application of LIF for studies of atomic species such as F, O, H, and Cl. The lowest energy of electronic transition of these species is in the vacuum ultraviolet, beyond the reach of commercially available tunable lasers.

Laser induced fluorescence measurements are characterized by very high sensitivities. Molecular species at a density as low as 10^6–10^7 cm^{-3} can be detected by LIF. The detection sensitivity decreases to a minimum of 10^{13} cm^{-3} for atomic species whose excitation usually requires simultaneous absorption of two photons.

The advantages of laser induced fluorescence include, in addition to the high sensitivities, selectivity (there is usually no detectable interference from other species) and the provision of temporal and three-dimensional spatial resolution. LIF overcomes the two major problems of the optical emission spectroscopy: the undefined source of the emitting species and the dependence of the intensity of the emission on plasma density. In addition, LIF provides a three-dimensional resolution, while OES has only a two-dimensional one. Laser induced fluorescence can be used also in the absence of plasma, while OES cannot. LIF can be more quantitative than OES because the energy of the exciting photons and their number can be controlled.

Because of its high sensitivity, laser induced fluorescence is particularly well suited to study the motion of ions, which are present at levels below 10^{11} cm^{-3}. Such studies have been carried out on Cl_2^+ and N_2^+ (see Donnelly [32]). LIF can also be used to measure velocity distribution of atoms and small molecules in plasma.

The disadvantage of the LIF technique is that it is only useful for those species with bound excited electronic states that can be reached by optically allowed transitions from the ground states. Another disadvantage of the LIF technique is the complexity of the required apparatus.

5.3.4 Summary

Optical plasma diagnostic methods are widely used for investigation of chemical processes in cold plasmas and plasma-surface interactions. Optical methods are used both for studying basic plasma phenomena and for analysis of processing plasmas. No single optical method can be regarded as a universal plasma diagnostic tool. Several complimentary techniques are required to provide detailed information on a variety of processing plasmas.

Of all optical techniques, optical emission is the easiest to implement and can provide qualitative data on concentrations of species. OES provides real time information that can be used to improve control of plasma processing techniques such as etching and deposition. It can also provide quantitative information on relative number densities as well as discharge kinetics and dynamics.

Optical absorption spectroscopy requires a somewhat more complex degree of instrumentation but can supply absolute concentration measurements and is capable of temporal and spatial resolution.

LIF is relatively complex and many species cannot be detected due to a lack of excited bound states at energies accessible to tunable dye lasers. Yet, it is by far the most sensitive technique and is uniquely suited to detection of trace species, including ions.

5.4 QUESTIONS

1. What determines the range of masses that can be analyzed by a quadrupole mass spectrometer and its resolution? What is the resolution of the QMS?
2. How is the transmission of the quadrupole mass analyzer affected by the mass of the analyzed ions? What is the result of this dependence?
3. Describe briefly the components of the QMS and the role of each of them.
4. What determines the choice of the detector for a quadrupole mass spectrometer?
5. Describe experimental arrangements for mass spectrometry analysis of a low-pressure chemical vapor deposition system operating at 0.1 torr and a plasma system operating at the same pressure. What are the differences between the two types of analyses and what kind of experimental arrangements are possible?
6. Discuss the differences between quadrupole mass spectrometric analyses of neutral and ion fluxes.
7. What are the differences between plasma diagnostics by double floating electrostatic probes and single Langmuir probe? What are the advantages of DFEP plasma diagnostics?
8. Figure 5-14 shows an IV trace obtained from a single Langmuir probe. The probe is cylindrical, 0.8 mm in diameter and 3 mm long. Calculate the electron temperature, electron density, and ion density of the plasma from which this probe characteristic was obtained.
9. a. What affects the accuracy of evaluation of plasma properties by electrostatic probe measurements?

   b. What can preclude the use of electrostatic probes for plasma diagnostics?
10. a. What are the differences between the experimental setups for optical emission spectroscopy and absorption spectroscopy?

    b. Which plasma species can be analyzed by each method?
11. a. What is actinometry and how can it be used for plasma analyses?

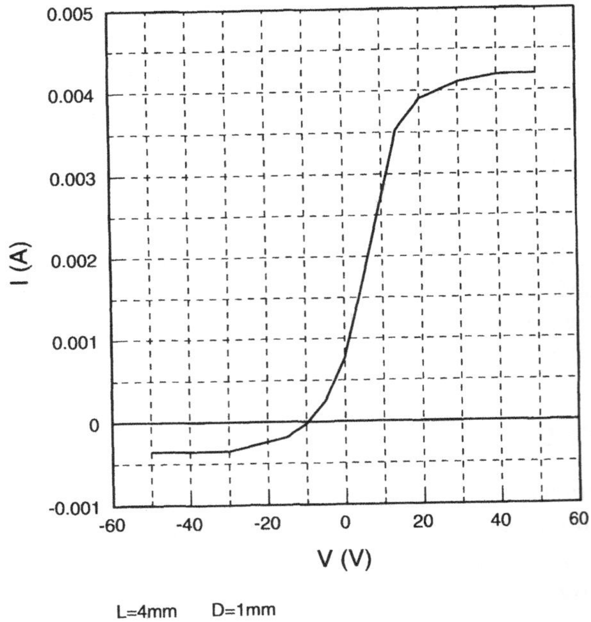

Fig. 5-14 I-V trace of single Langmuir probe for question 8.

b. What types of gases can be used for actinometry and what conditions do they have to satisfy?

12. What are the principles of LIF and what are its advantages and disadvantages for plasma diagnostics?

5.5 REFERENCES

[1] Huddlestone, R. H., and S. L. Leonard, eds., *Plasma Diagnostic Techniques*. New York: Academic Press, 1965.

[2] Loechte-Holtgreven, W., ed. *Plasma Diagnostics*. Amsterdam: North-Holland, 1968.

[3] Venugopalan, M., ed. *Reactions Under Plasma Conditions*, Vol. I. New York: Wiley-Interscience, 1971.

[4] Heald H., and C. B. Wharton, *Plasma Diagnostics with Microwaves*, p. 383. New York: Wiley-Interscience, 1965

[5] Hutchinson, I. H., *Principles of Plasma Diagnostics*. Cambridge: Cambridge University Press, 1987.

[6] Auciello, O., and D. L. Flamm, eds., *Plasma Diagnostic Techniques*. New York: Academic Press, 1989.

[7] Yasuda, H. K., *Plasma Polymerization*. New York: Academic Press, 1985.

[8] Graves, D. B., *AIChE J.*, 35: 1 (1989).

[9] Paul, W., H. P. Reinhard, and U. von Zahn, *Z. Phys.*, 152: 143 (1958).

[10] Dawson, P. H., ed., *Quadrupole Mass Spectrometry and Its Applications*. New York: Elsevier, 1976.

[11] Austin, W. E., In *Quadrupole Mass Spectrometry and Its Applications*, ed. P. H. Dawson, p. 121. New York: Elsevier, 1976.

[12] Vasile, M. J., and H. F. Dylla, In *Plasma Diagnostics*, eds. O. Auciello and D. L. Flamm, Vol. 1, p. 185. Washington, DC: Academic Press, 1989.

[13] Drinkwine, M. J., and D. Lichtman, In *Partial Pressure Analyzers and Analysis*, eds. N. Rey and R. Long, Jr., p. 24. Milwaukee: American Vacuum Society Monographs Series, 1989.

[14] Longway, P. A., R. D. Estes, and H. A. Weakliem, *J. Phys. Chem.*, 88: 73 (1984).

[15] Vasile, M. J., and G. Smolinsky, *Int. J. Mass Spectrom. Ion Phys.* 21: 264 (1976).

[16] Manory, R. R., A. Grill, U. Carmi, and R. Avni, *Plasma Chem. Plasma Process.*, 3: 235 (1983).

[17] Turban, G., Y. Catherine, and B. Grolleau, *Thin Solid Films* 67: 309 (1980).

[18] Coburn, J. W., In *Diamond and Diamond-like Films and Coatings, NATO-ASI Series B: Physics*, eds. R. E. Clausing, L. L. Horton, J. C. Angus, and P. Koidl, Vol. 266, p. 73. New York: Plenum Press, 1991.

[19] Drawin, H. W., In *Plasma Diagnostics*, eds. W. Loechte-Holgreven, p. 777. Amsterdam: North-Holland, 1968.

[20] Haller, I., *Appl. Phys. Lett.*, 37: 282 (1980).

[21] Catherine, Y., G. Turban, and B. Grolleau, *Plasma Chem. Plasma Process.* 2: 81 (1982).

[22] Mayo, N., U. Carmi, I. Rosenthal, R. Avni, R. R. Manory, and A. Grill, *J. Appl. Phys.*, 55: 4404 (1984).

[23] Manory, R. R., U. Carmi, R. Avni, and A. Grill, *Thin Solid Films* 156: 79 (1988).

[24] Carmi, U., A. Inspector, and R. Avni, *Plasma Chem. Plasma Process.*, 1: 233 (1981).

[25] Day, A. P., D. Field, Y. P. Soung, and D. F. Klemperer, *Semicond. Int.* 110 (November, 1989).

[26] Swift, J. D., and M. J. Schwar, *Electrical Probes for Plasma Diagnostics*. New York: Elsevier American, 1969.

[27] Chen, F. F., In *Plasma Diagnostic Techniques*, eds. Richard H. Hudlestone and Stanley L. Leonard, p. 113. New York: Academic Press, 1965.

[28] Chung, P. M., L. Talbot, and K. J. Touryan, *Electric Probes in Stationary and Flowing Plasmas*. Berlin: Springer-Verlag, 1975.

[29] Hershkowitz, N., In *Plasma Diagnostics,* eds. O. Auciello and D. L. Flamm, Vol. 1, p. 113. New York: Academic Press, 1989.

[30] Venugopalan M., and R. Avni, In *Thin Films from Free Atoms and Particles*, ed. Kenneth J. Klabunde, p. 59. London: Academic Press, 1985.

[31] Cabannes F., and J. Chapelle, In *Reactions Under Plasma Conditions*, ed. M. Venugopalan, Vol. I, p. 367. New York: Wiley-Interscience, 1971.

[32] Donnelly, V. M., In *Plasma Diagnostics*, eds. O. Auciello and D. L. Flamm, Vol. I, p. 1. Washington, DC: Academic Press, 1989

[33] Krogh, O., T. Widser, and B. Chapman, *J. Vac. Sci. Technol.* A 4: 1976 (1983).

[34] Gottscho, R. A., and T. A. Miller, *Pure & Appl. Chem.* 56: 189 (1984).

[35] Mogab, C. J., A. C. Adams, and D. L. Flamm, *J. Appl. Phys.* 49: 3796 (1979).

[36] Field, D., D. F. Klemperer, Y. P. Soung, P. V. Denis, and A. P. Webb, In *Proc. 9th Int. Symp. on Plasma Chemistry*, ed. R. d'Agostino, p. 522. Italy: Pugnochioso, IUPAC, 1989.

[37] Coburn, J. W., and M. Chen, *J. Appl. Phys.* 51: 3134 (1980).

[38] D'Agostino, R., and F. Illuzi, In *Deposition, Treatment and Etching of Polymers*, ed. R. d'Agostino, p. 95. New York: Academic Press, 1990.

[39] D'Agostino, R., F. Cramorosa, S. DeBenedictis, and G. Ferraro, *J. Appl. Phys.* 52: 1259 (1981).

[40] Miller, T. A., *Plasma Chem. Plasma Process*, 1: 3 (1981).

6

Cold Plasma Processes for Surface Modification

6.1 INTRODUCTION

The use of cold plasmas for materials processing has gained acceptance in a large number of technologies, from metallurgy to manufacturing of computer chips, and covers a wide variety of materials, including metals, semiconductors, and polymers. Most applications of cold plasmas based on plasma enhanced chemical processing focus on treating solid surfaces, film deposition, modification of surfaces, and etching of surface layers. Cold plasmas are also used to synthesize gases or liquids as final products.

Plasma enhanced chemical processing (*PECP*) takes advantage of the high-energy electrons present in glow discharge to dissociate and ionize the molecules of the gas to form chemically reactive radicals and ions. Because in such conditions thermal energy is not needed to break chemical bonds, reactions can be promoted at low temperatures (< 200 °C). These relatively low temperatures are providing one of the main benefits of processing with cold plasmas. With PECP it is possible to process substrates that do not have the thermal stability necessary to withstand processing at higher temperatures. Low processing temperature may be imposed by the the substrate or by the composition of the depositing film (e.g., polymer). Low temperatures may also be required to preserve properties achieved by prior processes, for example, heat treated metals. In addition, treatment at lower temperatures possible in the plasma is critical when one has to limit diffusion, or grain growth, especially in microelectronics where very small feature sizes and shallow junctions are involved.

All plasma enhanced chemical processes are the result of chemical reactions, the main difference between the various processes being the state of the final product. For deposition of films, the product of the reactions has to be nonvolatile and deposit on the substrate. In some surface modification applications, such as nitriding or oxidation, the gas is dissociated in the plasma, forming radicals that react with the surface to form a nonvolatile product that remains on the surface and changes the characteristics of the surface layers. For cleaning, ashing, or etching, the products of the reactions between plasma species and substrate have to be volatile to remove material from the exposed surface.

Cold plasmas can deliver a uniform flux of energetic particles (e.g., positive ions, neutrals, metastable species, electrons, and photons) whose energy and flux can be controlled. The substrates are therefore subjected to bombardment with energetic particles, which imparts both physical and chemical components to the plasma treated surfaces. The bombardment of the surface by the ions accelerated through the plasma sheath can affect the processes occurring at the surface and the properties of the processed surface or deposited film. The physical component can induce directionality during film etching, or it can alter the properties of the deposited film by promoting surface reaction steps. The combination of the physical processes with the purely chemical reactions yields process rates and material properties unattainable with either process separately.

The use of plasma processing has been expanded due to the necessity to avoid problems associated with the disposal of chemicals used in alternative wet processing techniques.

A summary of applications of cold plasmas is presented in Table 6-1. The applications are classified according to the type of feed gas, the product of surface reactions, and the intensity of ion bombardment.

The next chapters will describe various applications of cold plasmas. It is not the intention of this book to describe in detail the different types of applications, nor is it the intention of the author to describe each possible application of cold plasma technology, or to dedicate equal amounts of discussion to the selected

TABLE 6-1 Classification of Plasma Applications (after [1], reprinted with permission)

	Inert Gas Plasma	Reactive Gas Plasma	
		Nonvolatile surface product	Volatile surface product
Low ion bombardment	Plasma cleaning	Plasma oxidation, surface activation, surface hardening, polymerization, film deposition	Surface activation, plasma ashing, plasma etching
High ion bombardment (high negative bias)	Sputtering, sputter etching	Reactive sputter etching	Reactive ion etching

applications described in the following chapters. Such details can be found in reviews and monographs, which will be referenced for the specific applications, or in current publications in scientific journals.

The present and next two chapters consist of a selection of applications of cold plasmas, and typical cases will be discussed alongside the basic principles of these applications. The chapters are organized according to the state of the end product of the main groups of processes: surface modification, deposition of coatings, and etching. Taking into consideration the relative novelty of the field and the amount of current research and development ongoing in the field, the reader is advised to consult the most recently published literature in scientific journals.

Surface modification processes discussed in the present chapter are procedures that change the properties of the outer layers of a solid, in a region close to its surface. This category of plasma processing generally refers to activation, surface cleaning, oxidation, and hardening. The plasma surface modification processing is implemented in manufacturing of microelectronic components, magnetic and optical recording, mechanical tools, treatment of polymers, and so on.

6.2 SURFACE MODIFICATION OF POLYMERS

The use of cold plasmas for surface activation of polymers was suggested already in 1968 by Beauchamp and Buttrill [2]. It is generally difficult to adhesively bond polymers to themselves or other materials. Polymers such as fluoropolymers, characterized by high chemical inertness, thermal stability, and electrical insulating properties, also have very low surface tension, which causes negligible bonding to other materials and especially metals. Plasma surface treatments allow the modification of the surface characteristics of polymers to obtain improved bonding, without affecting bulk properties. The methods employed to modify properties of polymeric surfaces include *surfaces activation, deposition*, and *grafting*. We shall first discuss surface activation, which is especially important for polymeric materials.

Exposure of polymers to suitable plasmas can cause chemical and physical changes to their surface or near-surface layers. These changes produce more reactive surfaces and affect wetting properties, cross-linking, and molecular weight. Surfaces can thus be engineered to achieve the necessary properties, such as wettability, adhesion, barrier protection, material selectivity, and even biocompatibility [3–6]. *Surface activation* by plasma treatment of a polymer refers to the modification of the properties of its surface by exposure to a nonpolymer-forming plasma [7, 8].

The surface activation of polymers is based on the interaction of their surfaces with the energetic plasma particles. Plasma activation is performed in plasmas excited in gases that do not polymerize even in a plasma, such as oxygen, nitrogen, nitrogen oxides, ammonia, or inert gases. The bombardment of the polymeric surfaces with the energetic particles breaks the covalent bonds on the surface of the bombarded polymer and leads to the formation of surface radicals on the treated polymer. These surface radicals, in turn, react with the active species of the plasma to form various active chemical functional groups on the surface of the substrate. Surface contaminants and weakly bounded polymer layers can dissociate into volatile by-products that are removed by pumping from the reactor.

Oxygen and nitrogen plasma exposures for a few to a few tens of minutes can make hydrophilic the surfaces of most polymers, including polyolefines, polyesters, polyamides, and fluoropolymers [9]. The active plasma species attack the polymer surfaces and cause the incorporation of hydrophilic groups, such as carbonyl, carboxyl, hydroxyl, and amino groups, into the surface of the polymers. Inagaki has reported that radical species, rather than ionic species and electrons, play an important role in the hydrophilic modification [9].

Treatment of various types of materials in an oxygen plasma is one of the methods available to improve the adhesion and wettability of surfaces. The plasma treatment causes the introduction of polar groups onto the surface of the polymer resulting in adhesion improvement [10]. Oxygen species can themselves be incorporated in the surface of the treated polymer. However, polymer degradation and surface cross-linking also take place in oxygen plasmas. Polyethylene surfaces modified by oxygen radicals isolated from the plasma were found to possess almost the same functionality as those modified by direct exposure to oxygen plasma. Both treatments produced carbonyl and carboxyl groups on the surface of the polymer, and the surfaces treated in the oxygen plasma also contained a very small amount (0.02%) of hydroxyl groups. The plasma treated surfaces showed good adhesion and printing abilities [9].

X-ray photoelectrons spectroscopy (XPS) analysis of Teflon® PFA (perfluoro alkoxyl) treated in an oxygen plasma has shown that the O_{1s} peak disappeared after 10 minutes of exposure to the plasma. The ratio of the F_{1s}/C_{1s} peaks remained almost unchanged as compared to the untreated sample [10]. The results showed that the plasma treatment preferentially broke the carbon-oxygen bonds in Teflon® PFA, depleting the surface oxygen. It has also been observed that the plasma treatment hardly incorporated oxygen into the surface of the Teflon®, but caused the formation of peroxy radicals of $-CF_2CF_2OO^{\bullet}$ and $-CF_2CF(OO^{\bullet})CF-$ types, the amount of which was strongly dependent on the treatment conditions [10].

As a result of 10 minutes of plasma treatment of polyethylene in NO and NO_2 plasmas, both oxygen and nitrogen were incorporated into the polyethylene surface. Hydroxyl, carbonyl, and carboxylic acid group were formed as oxygen functionalities and amino groups were formed as nitrogen functionalities. The concentration of the functionalities was found to be three to seven times higher in the NO plasma treated than in the NO_2 plasma treated polyethylene [9]. The treatment made the surfaces hydrophilic, as determined by measuring the advancing contact angle of water and the surface energy.

Nitrogen species from nitrogen or ammonia plasmas are often incorporated in the surface of polymers treated in plasmas containing amine and amide-type groups. Treatment of polytetrafluoroethylene (PTFE) in a nitrogen plasma was found to increase its hydrophilicity, as indicated by reduced contact angle to water, and increased the adhesion strength of plasma polymerized hexamethyldisiloxane (PHMDS) films to PTFE [11].

Low-temperature RF plasmas of ammonia or mixtures of nitrogen with hydrogen have been used to modify surfaces of polymers such as polypropylene, PTFE, polycarbonate, polyurethane, polyvinyl chloride (PVC), and polymethylmethacrylate (PMMA). Amino groups formed in the plasma attach to the polymers rendering them more wettable. The degree of amino attachment to the polymer was found to be a function of the plasma parameters [12].

Plasma activation can be used also for the opposite conversion of hydrophilic surfaces to hydrophobic surfaces. This can be done with fluorocarbon plasmas, which cause the replacement of surface C-H bonds with C-F bonds or add CF_2 and CF_3 groups to the surface of the polymer [13].

Exposure of a polymer to plasmas of inert gases generally results in physical ablation and formation of free radicals of the polymer. Ablation takes place by ion sputtering and impinging of energetic neutral species. The ablated fragments can be activated in the plasma and reincorporated in the surface of the polymer or be redeposited on the surface as a highly cross-linked layer. Porous membranes for reverse osmosis were produced in this way by modifying the surface of a porous acrylonitrile film in a helium or an oxygen plasma [8]. UV radiation emitted in the plasma also causes formation of polymer-free radicals. Infrared spectroscopy studies of structural changes in polyimide during RF argon ion sputtering have shown the formation of a damaged layer, having a graphiticlike structure, at the polyimide surface. It has been found that this layer attains a constant thickness as sputter etching progresses [14].

Surface activation of PTFE can be obtained by treatment in an acetylene plasma, resulting in the formation of a rough, cross-linked surface layer. Sputter cleaning the surface by an inert gas plasma can improve adhesion of PTFE to copper, by roughening the surface of the polymer through breaking its polymer chain. Improved adhesion and wettability of PTFE has been obtained by sputter etching the surface in a RF argon plasma. The plasma treatment caused the formation of cones 1 μm in height with a surface density of 10^9 cm^{-2} and formation of $-C-O$ bonds. The changes in the wettability of the treated surface were attributed by the authors to physical effects, while adhesion improvement was attributed to both physical and chemical changes [3]. However, the chemical changes caused by the plasma treatment could also affect the wettability.

Addition of oxygen to a helium or an argon plasma causes the reaction between oxygen and the broken polymer chains, producing volatile compounds and leaving an activated surface. The adhesion strength showed a threefold strength increase after treating in an oxygen-containing plasma as compared to the treatment in the noble gas plasma [3].

In addition to activation of polymer surfaces, cold plasma processing is used to produce chemical modification of the surfaces by deposition and grafting. In deposition, plasmas of complex gases such as methane or other volatile monomers are used because these gases can polymerize in the plasma. The dissociation in the plasma of the monomer precursors creates various active fragments that react to form a film on the surface of the substrate, whose properties are different from the properties of the bulk polymer. Although deposition is mentioned here as a means of modifying the surface of an existing polymer material to make it more active, the same process is used to grow new polymeric films and will be discussed in detail in Sec. 7.1.

Several approaches for surface modifications of polymers are illustrated in Fig. 6-1 [8]. In this figure the modifications are divided into two categories according to the substrate morphology, that is, porous or nonporous. Plasma

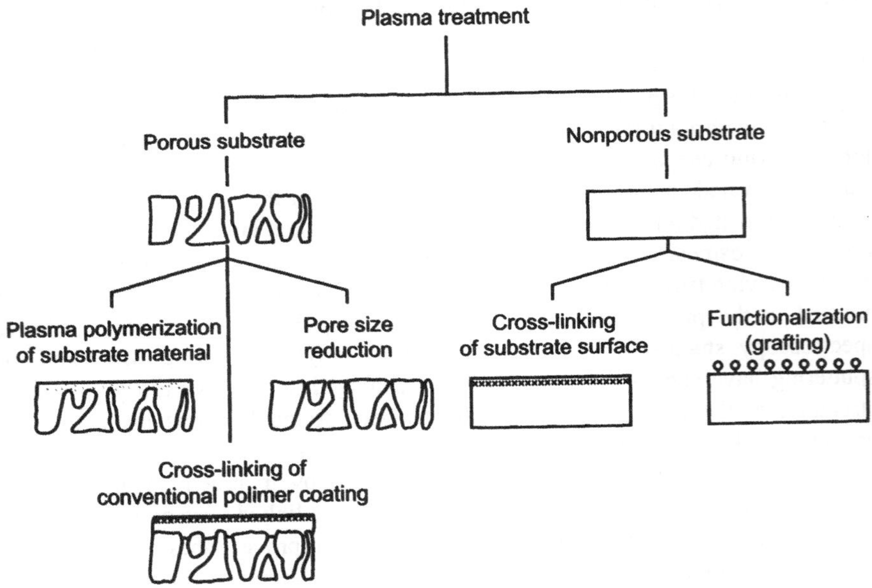

Fig. 6-1 Surface modification of polymers by plasma treatment (from [8], reprinted with permission from *J. Membr. Sci.*, vol. 46, p. 1, 1989).

treatment of microporous membranes is in fact a plasma polymerization of gases sputtered or ablated from the substrate. This type of plasma treatment may close completely the porosity of the substrate. For example, membranes for reverse osmosis have been prepared by treating porous acrylonitrile copolymer films with helium or oxygen plasmas [8]. If the plasma treatment time is short, the pores of the modified substrate are not completely closed by polymerization of the ablated material.

A porous polymer can also be modified by depositing a conventional polymer on its surface and crosslinking it in the plasma. Gas separation membranes have been prepared by this method by depositing a silicon polymer on a porous substrate and then treating it for a short time in an argon plasma.

Two types of plasma modifications of nonporous substrates are illustrated in Fig. 6-1. One process, also called CASING (cross-linking by activated species in inert gas plasma), causes cross-linking of the polymer surface or of a film priorly deposited on the surface, by bombardment with energetic plasma species. CASING of a poly(dimethylsiloxane) in an argon plasma has been used to create a semipermeable membrane by causing the reduction of oxygen permeability through the film [8]. Plasma cross-linking has also been reported for polyethylene and polyvinylchloride [15].

Surface modification of polymers can also be produced by functionalizing the surface through *plasma induced grafting*, a hybrid form of plasma activation and conventional chemistry. In this process, a plasma excited in an inert gas, such as argon or helium, activates the surface by creating free radicals on the surface of the sample. After plasma activation and before exposure to the atmosphere, the material's surface is exposed to an unsaturated monomer, causing a polymer layer to be grafted onto the activated substrate surface.

Functionalization of the polymer surface can be achieved by grafting of functional groups created by the plasma. Thus plasmas of ammonia or mixtures of nitrogen with oxygen have been used to create amino groups that then attached to various polymer surfaces, without altering the properties of the polymer bulk. Allyl amine and allyl alcohol plasmas have been used to deposit reactive NH_2 or -OH groups for subsequent chemical immobilization of polyethylene oxide, which leads to reduction of protein adsorption. Such surface modifications have been used for biomaterial applications. Ammonia or nitrogen plasma have been used to produce amine groups for direct ionic binding of proteins and cells, and oxygen plasmas were applied for subsequent covalent binding of polysaccharides or of enzymes. Acrylic acid and methyl methacrylate are most easily grafted to treated polymer surfaces [6, 15–17], as illustrated in Fig. 6-2.

Figure 6-2 also illustrates other types of plasma reactions used for surface activation of polymers. In this example, oxidation in an oxygen plasma was used to remove surface contaminants from the fibers of a polyethylene fiber composite and to incorporate polar groups on the surfaces of the fibers. An RF plasma caused the formation of chemically active oxygen species, which reacted with the surface of the polymer. After exposure to the plasma, hydroxyl and carboxyl acid groups were formed on the surface of the fiber, and after 1 minute of plasma treatment, a 15% increase in surface oxidation was obtained. This surface activation treatment enhanced the chemical bonding of the fibers to the resin matrix of the composite and resulted in a composite structure with higher flexural modulus and shear strength than structures made with untreated fibers [6]. A similar improvement of adhesion was performed by treating Kevlar® fibers in an ammonia plasma. A composite made from the treated fibers and epoxy had better mechanical properties and improved resistance to moisture penetration [15]. Polymers can also be plasma treated to increase their stability by causing cross-linking at the surface.

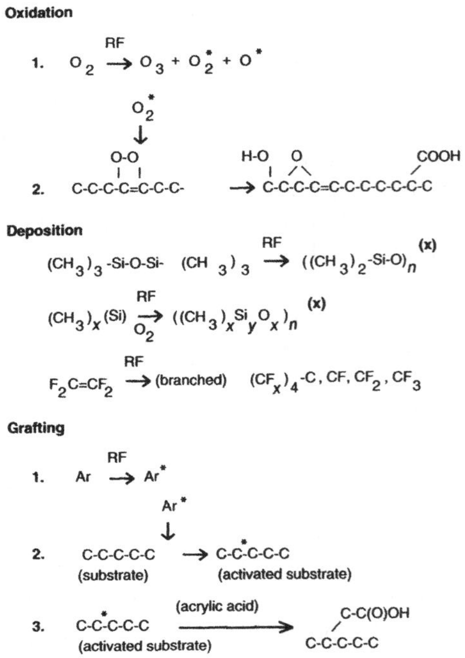

Fig. 6-2 Examples of surface activation of polymers by plasma treatment (from [6], reprinted with permission from *Research & Development Magazine*, July 1990, by Cahners Publishing Company). (**x**): Since the structure of the plasma polymer is not well defined, the formula of the product may be misleading: here the index *n* indicates only a large molecule formed from fragments of the original monomer.

Plasma treatments can improve the adhesive bonding characteristics of engineering plastics, such as polycarbonates, polyacetals, aromatic esters, imides, amides, and liquid crystal polymers. Following plasma treatment, the adhesive bond often becomes stronger than the substrate material [6]. Tetramethyloxysilane, dimethyldimethoxysilane, and hexamethyldisilane were polymerized on titanium to improve adhesion of alkyd paints to the metal [18].

As mentioned before, especially difficult to bond to are fluoropolymers (e.g., Teflon®) because of their inert characteristics. A copolymer of tetrafluoroethylene and ethylene (DuPont Tefzel®) treated with an oxygen-sulphur hexafluoride ($O_2 + SF_6$) plasma increased threefold its lap shear strength as compared to untreated surfaces [3]. Other examples of plasma activation for adhesion improvement are presented in Table 6-2.

Surfaces of biomaterial polymers employed in biomedical products are among the materials modified by plasma treatment. Biomaterials are materials used in contact with body fluids, but foreign to the body. A wide variety of biomaterials are used for in-vivo, extracorporeal, and topical medical implants; passive and bioreactive medical devices; and drug delivery and toxin removal systems. They are also used as biosensors, or in industrial bioprocesses, as cell carriers or enzyme supports, and for separation and purification systems [4]. In all cases, biomaterials contact biological environments that contain biomolecules, such as proteins or living cells, and the surface of the materials must have qualities

TABLE 6-2 Applications of Surface Modifications of Polymers for Adhesion Improvement (after [15])

Substrate	Plasma gas	Plasma conditions
Polyaramid (Kevlar®)	Ammonia	
Polyaramid	Hexamethyldisiloxane	2.45 GHz, 0.2-1 torr
Polyethylene, polyvinylidenefluoride, polytetrafluoroethylene	Acetylene	50 Hz, capacitively coupled, 0.1–0.7 torr, 4.7–42 sccm

that make them compatible with live environment and allow them to interact favorably with live tissues, fluids, and proteins.

The interaction of the biomaterial with living cells or proteins might have to be modified, either to reduce or to enhance that interaction according to the application. Such modifications can be obtained by plasma treatments. These interactions are affected by the biomaterial's surface properties, such as [4]:

1. Chemical composition (polar-apolar, acid-base, H-bonding, ionic charges)
2. Surface microstructure (polymer chain ends, loops and their flexibility)

TABLE 6-3 Biomedical Applications of Cold Plasmas (from [4], reprinted by permission of John Wiley & Sons, Inc.)

A. *Plasma Treatment* (Etching)
   1. Clean.
   2. Sterilize.
   3. Cross-link surface molecules.

B. *Plasma Treatment (Etching) and Plasma Polymerization (Deposition)*
   1. Form barrier film:
      a. Protective coating
      b. Insulating coating
      c. Reduce absorption from environment
      d. Reduce release rate of leachables
      e. Control drug delivery rate
   2. Modify protein and cell interactions:
      a. Improve "biocompatibility"
      b. Promote selective protein adsorption
      c. Enhance cell adhesion
      d. Improve cell culture surfaces
      e. Provide nonfouling surfaces
      f. Reduce surface friction
   3. Provide reactive sites:
      a. For grafting or polymerizing polymers
      b. For immobilizing biomolecules

3. Topography (roughness, porosity, imperfections, gas microbubbles)
4. Domains (distribution of any of the foregoing in the surface)

A variety of plasma treatments have been used to provide biomaterials with reactive groups for subsequent bonding of biomolecules. A list of plasma treatment of biomaterial polymers for different applications is presented in Table 6-3.

Surfaces of materials used in vascular prostheses, blood oxygenators, catheters, and contact lenses and for cell culturing containers have been treated in cold plasmas to bind heparin (a blood anticlotting agent), thereby retarding blood coagulations when the heparin-coated polymer was contacted with blood [6]. Polystyrene, with its optical clarity, low cost, and durability, is often used to make disposable culture dishes. However, living cells do not readily spread and grow on untreated polystyrene. Surface activation in an oxygen plasma alters the surface chemistry of polystyrene and allows the cells to spread and significantly increase their cloning efficiency [6].

6.3 SURFACE CLEANING AND ASHING

6.3.1 Cleaning

In advanced VLSI manufacturing, cleanliness of the wafers is of utmost importance. The goal of any overall wafer cleaning process is to remove from the silicon surface photoresist, residual organics, residual metallic contaminants, and native oxide or oxides formed on the surface during the cleaning itself. Several factors favor the dry plasma cleaning as replacement for wet cleaning of wafers. The VLSI structures include deep trenches with high-aspect ratios (aspect ratio = depth / width) and many problems are associated with the wet processing of high-aspect ratio structures. Complications arise from the difficulty of getting liquid into and out of small openings. In addition, wet cleaning is not always compatible with integrated chip processing, which requires complete removal of trace contaminants from the wafer. This can only be done by volatilizing the residual contaminants and removing them from the gas phase.

The control of particles during wafer treatment in liquid chemicals is also very difficult, especially when at least part of the contamination is caused by the wet cleaning medium. Metallic contaminants, for example, might be electrochemically plated on silicon surfaces from NH_4OH- and/or HF-containing solutions. Organic contaminants may originate from containers and utensils used in wet cleaning operations. Conversion to "all-dry" cleaning can reduce these problems significantly since particles in gas are easier to control than those in liquids [19].

The volatilization of trace amounts of organic contaminants can be accomplished fairly easily in a plasma. For this purpose, the cleaning plasma contains an oxidizing gas that converts organic contaminants from the surface of the sample into gaseous oxides of carbon and other gaseous reaction product. A remote oxygen plasma can be used to prevent inducing damage in the substrate during cleaning [19].

A surface exposed to the plasma is bombarded with energetic ions. The kinetic energy of the ions can cause sputtering, ion implantation, heating, and migration. The energy of physical adsorption of gases on the substrates is 0.1–0.5 eV, and that of chemisorbed gases is 1–8 eV [20]. The energy of striking ions should therefore be larger than these values to cause desorption of the gases for cleaning the surface.

Cleaning of a silicon surface can be performed in plasmas of inert gases and/or hydrogen. Such a plasma of argon + hydrogen has been used to clean the surface of silicon wafers prior to the deposition of epitaxial silicon films by high-vacuum chemical vapor deposition (CVD) [21]. It was found that, as a result of the plasma cleaning, the epitaxy temperature could be lowered to 500 °C. Cleaning by sputtering in pure argon plasma requires substrate biasing, which can often result in substrate damage. Superposing the chemical effect of hydrogen atoms, by cleaning in hydrogen or argon + hydrogen plasma, at zero bias is preferable because substrate damage can be avoided in the absence of ion bombardment [22]. Using an ECR plasma of hydrogen, it was found that the cleaning in hydrogen plasma is as effective for unbiased as for biased substrate. A helium plasma was found to be less effective in cleaning the silicon surface, indicating that the chemical reaction between the native oxide and chemically active hydrogen is responsible for the cleaning rather than just sputtering by pure helium ions [23].

The removal of metallic contaminants from the surface of silicon wafers is very complex because one must deal with a variety of metals, each with different reactivity, segregation coefficients, and diffusion in silicon and silicon oxide. The metal removal must be carried out initially at low temperatures to prevent metals from being driven through diffusion into the silicon substrate before they can be removed from the surface. However, at low temperatures, the metallic species of concern or their compounds formed in the plasma are not volatile. Therefore, as far as temperature is concerned, gas-phase metal removal must typically be a two-stage process. In the first stage, metallic species are converted into complexes that trap the metal at the surface and prevent its diffusion into the silicon wafer. This can be done in a suitable gas mixture at a temperature below 200 °C. The complexes are next volatilized by increasing the temperature of the substrate and reducing the pressure so that sufficient vapor pressure is achieved [19].

Volatile compounds of metallic species can be produced with halogens. Among the halogens, chlorine reacts with almost all the metals of interest and generally produces species more volatile than other halogen compounds. A chlorine-based plasma can therefore be used for surface cleaning from metallic contaminants.

The complexity and intricacies of a plasma cleaning process for VLSI fabrication are described next. A dry-cleaning procedure was designed to dry clean Si wafers in a remote plasma, by exposure to the downstream effluent of a microwave plasma of suitable composition [19]. The procedure consisted of a sequence of seven steps preceding the formation of the gate oxide of the electronic devices. The remote plasma configuration was chosen to minimize substrate damage. The details of the plasma cleaning procedures are shown in Table 6-4.

TABLE 6-4 Remote Plasma Cleaning Process (from [24])

STEP	Gas Composition	T (°C)	p (torr)	Etch Rate (nm/min)	Process Time (sec)
1. Resist strip (1000 nm)	O_2	200	1	650	120
2. Residual oxide (1.4 nm) etch	NF_3 1% H_2 2% Ar 97%	200	1	2.5	40
3. Fine organics removal/oxide regrowth (1.3 nm)	O_2	200	1	—	30
4. Metal complexing	HCl 10% Ar 90%	200	1	Si etch 30 nm	30
5. Residual oxide etch/partial volatilization of metal compounds	NF_3 1% H_2 2% Ar 97%	200	1	2.5	40
	Temperature Ramp*				
6. Final volatilization of metal compounds	vacuum anneal	750	0.02	—	120
7. Protective oxide regrowth (2 nm)	O_2	750	1	—	25

* Time of temperature ramp depends on heating elements used.

Steps 1–3 in Table 6-4 were required for the complete removal of organic materials from the surface. Ultrathin silicon oxide formed during the first step was removed in the second step. Hydrogen was added to the plasma mixture to scavenge fluorine and prevent excessive etching of the silicon substrate. Steps 4–6 were designed to volatilize metallic contaminants [24]. Metals were converted to chloride, which were then volatilized by heating to higher temperature. During step 5 the residual oxide was etched off, and some chloride trapped in it was removed. The anneal at 750 °C in step 6 was aimed at final removal of metallic contaminants, which were in the form of chlorides at this stage. An important consideration in the seven-step procedure was to keep the temperatures low at the initial steps to prevent diffusion of metals into the silicon before forming and evaporating the halide complexes [24].

Crystal quality of GaAs grown by metalorganic chemical vapor deposition on silicon has been shown to be improved when the substrate was cleaned in-situ in the reactor with a hydrogen + arsine (53 sccm + 2 sccm) plasma at 450 °C prior to film deposition. The critical step that controlled the interface cleanliness was the substitution of hydrogen atoms, passivating the Si surface after an ex-situ clean by HF dip, with As atoms. Reduction of the substrate temperature with the plasma cleaning process ensured very low kinetics of thermal desorption of hydrogen atoms, preventing reoxidation of the exposed Si surface and providing a fully As passivated surface [25].

Plasma cleaning of polymeric surfaces is an important step in many processes used in the electronics industry. Polyimide and photoresist surfaces are exposed to plasmas to oxidize carbon residues or remove thin superficial organic films [26]. Such a process can in certain conditions damage the polymer film. Cracking of polyimide films has been observed to occur during oxygen plasma treatment for removal of organic debris, at ion energies as low as 50 eV. The cracking of the polyimide was found to be a threshold phenomenon occurring after accumulation of sufficient exposure, which is cumulative in time [26]. Debris removal times have been observed to be sharply reduced as the pressure in the plasma was decreased. A decrease in pressure from 30 to 10 mtorr in a plasma of Ar/N_2O was found to decrease the debris cleaning time from 30–40 min to 10 min. The ratio between cracking threshold time and debris clean up time was found to increase with increasing ion density [26].

6.3.2 Ashing

Analysis of organic materials is often done after modifying or breaking them up by a process called *ashing*. Ashing is also used for removal of organic fragments from inorganic samples. Wet ashing is performed with oxidizing acid solutions, such as nitric or perchloric acid. The wet process can, however, destroy the physical structure of the sample as, for example, is the case for physiological material, and can also interfere with analysis of trace elements. Oxygen plasma, which contains atomic neutral or ionic oxygen, overcomes these disadvantages and proves to be a successful tool for ashing organic samples. Plasma ashing has been used since the late 1950s in a large spectrum of fields, from coal science to cell biology [27, 28].

Plasma ashing can be performed at temperatures varying between room and 300 °C. The process is performed in electrodeless discharges for [27]

- Dry ashing of organic or biological materials
- Removal of organic matrices, leaving inorganic residual structures for subsequent analysis
- Production of gaseous products to be analyzed by suitable methods
- Cleaning of surfaces from organic contaminants

Electrodeless plasmas are used to reduce to a minimum physical sputtering of the sample. Plasma ashing can be used on powders, flat sheets, microtoned sections, films, or porous materials.

While most of the ashing procedures are done in oxygen plasma, hydrogen plasma is also employed sometimes for ashing of carbonaceous materials, hydrocarbon, or protein samples. Sample degradation (ashing) in hydrogen plasma is due to the reaction of atomic hydrogen with oxygen-, carbon-, or nitrogen-containing fragments of the sample.

Another widely used application of plasma ashing is in photoresist stripping. At each stage of the fabrication of integrated circuits, the wafers are covered with a layer of photoresist. The photoresists are photosensitive organic polymers used in the photolithographic technique for device masking during fabrication. The photoresist is exposed to light through a mask and subsequently developed to create a pattern of areas free of photoresist. The exposed areas are etched or coated with another layer, while the remaining photoresist protects the rest of the wafer from the process. After the processing step, the photoresist has to be removed to prepare the wafer for the following fabrication steps. The removal of the photoresist, also called *resist stripping*, can be done in many cases by a wet procedure with chemical solutions. Wet stripping has, however, the disadvantage of requiring the handling and disposal of large volumes of either hydrocarbon solvents or strongly oxidizing acids and can leave traces of photoresist on the wafer that have to be completely removed. Removal of these residuals and better stripping can be done by ashing in a cold plasma [29].

Ashing of photoresist in an oxygen plasma was among the first applications of plasma chemistry in the 1960s. The organic photoresists consist mainly of carbon and hydrogen atoms. The photoresists do not react with molecular oxygen at room temperature. The chemically active atomic oxygen produced in an oxygen plasma reacts easily with the photoresist at room temperature to form volatile products, such as carbon monoxide, carbon dioxide, and water vapor. The volatile products are pumped out from the reactor together with the unreacted feed gases, as indicated in Fig. 6-3.

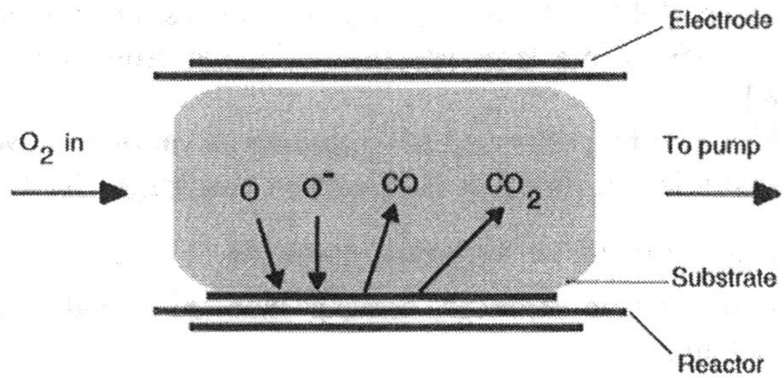

Fig. 6-3 Diagram of plasma ashing.

It is believed that plasma assisted resist stripping proceeds similar to thermal oxidation of polymers. The supply of active species in the plasma, however, enhances several steps that are often rate limiting [30]. The first step in resist stripping is oxidation of the polymer surface through oxygen atoms, which have been shown to be the most important reactive species in an oxygen plasma, and the etching rate has been found to correlate directly with the concentration of oxygen atoms. The effects of other oxygen species depend on the particular material being etched. In a plasma, a continuous supply of oxygen atoms, which rapidly abstract hydrogen, arrive at the etching surface. In addition, ion bombardment can create surface radical species and can aid in product desorption. Depending on the photoresist polymer, electron and photon bombardment may also cause radical formation and lead to subsequent reactions [30].

In stripping applications, the primary objective is the rapid removal of resist, without etching or damaging the substrate. For high rate stripping, the supply of the active species is rate limiting. The process is therefore usually performed at relatively high pressures, where ion energy is low and the reactant supply is high. Most resist stripping is performed in plasmas of mixtures of oxygen with either CF_4 or SF_6. Fluorine atoms are efficient at abstracting hydrogen in the first step of the oxidation reaction. In addition, the presence of fluorine atoms in the plasma enhances the dissociation of the oxygen, further increasing the ashing rates [31]. At low concentrations of CF_4 small amounts of excited fluorine atoms, F^*, assist the etching process. At high CF_4 concentrations the etching reaction competes with polymerization of tetrafluoromethane and fluorine passivation of the surface.

Etching rates during resist stripping increase with increasing temperature. The maximum usable temperature is limited by the decomposition temperature of the polymer. Resist stripping is typically performed at ~ 100 °C [30].

While resist stripping is usually performed in inductively coupled barrel reactors, introduced in 1970 by Bersin [32], several types of plasma reactors have been used for resist stripping, as shown in Table 6-5. In a parallel plate reactor the substrates are positioned on the grounded electrode for resist stripping and the substrates are thus exposed only to bombardment with ions of low energies limited by the plasma potential. To eliminate surface ion bombardment and reduce radiation damage, stripping can be performed in remote plasma reactors, downstream from the plasma zone, as for example, in the afterglow of a microwave oxygen plasma [33]. In such a configuration photon bombardment of the substrate is also

TABLE 6-5 Etching Rates of Resist Materials in Oxygen Plasmas in a Variety of Reactors (from [30], reprinted with permission)

Reactor Type	Pressure (torr)	Temp. (°C)	Resist	Etch Rate (nm/min)
(a) *RF plasmas*				
Downstream	1.0	100	KFTR	150
Barrel reactor with Faraday cage	1.0	150		250
Parallel plate	1.0	150	Kodak 747	350
Plasma mode	1.0	100	Kodak 747	140
Parallel plate (RIE mode)	0.02	40	AZ 2450	480
Magnetron RIE	0.05		Polyimide	2500
(b) *Microwave*				
Standard	0.02		HPR 204	270
Downstream	4.5	135	AZ 1370	350
	4.5	160	AZ 1370	1000

avoided. In microwave plasmas electron and ion densities are at least one order of magnitude higher than in RF plasmas. The increased ion density results in a higher degree of dissociation and thus significantly higher etch rates. A comparison of resist stripping rates in the different reactor configurations is shown in Table 6-5.

Plasma ashing in oxygen plasma does not remove inorganic contaminants from the surface of the processed wafer. The photoresist has to be therefore very clean, with minimum inorganic residuals. Plasma ashing in plasmas of oxygen and carbon tetrafluoride is also used in electronic packaging to remove organic smear formed by mechanical drilling of printed circuit boards to form electrical interconnection [13].

6.4 OXIDATION

The use of oxygen plasmas for growing oxide films on metal or semiconductor surfaces is one of the oldest techniques for surface modifications based on cold plasmas. In the VLSI technology one of the main problems is the reduction of process temperatures. Plasma oxidation of silicon, which can be performed at temperatures as low as 300–500 °C, offers distinct advantages over thermal oxidation, which is performed at much higher temperatures. Among these advantages are

- Reduced diffusion of heavy metals into the silicon.
- Possibility of oxidation after metallization; higher temperatures would cause melting of some of the metallic conductors.
- Possibility of oxidation of hydrogenated amorphous silicon, with reduced hydrogen loss that would otherwise occur and change the properties of the material.
- Prevention of wafer warpage as a result of reduced thermal stresses.

A summary of the reactions occurring in an oxygen plasma was given by Bell [34] (see Table 3-4 in Chapter 3). Among the more important reactions is charge transfer, in which a positive ion captures an electron from a neutral molecule. Atomic oxygen ions rapidly undergo charge exchange with oxygen molecules to produce molecular ions and neutral atomic radicals. In a typical glow discharge, the ratio of molecular ions to atomic ions is about 30:1 [35]. The degree of dissociation, which produces active oxygen atoms ranges from a fraction of a percent up to 20%. A typical plasma used in microelectronic processing may have only 1 to 2% oxygen atoms, with the dissociation increasing at higher power and lower pressures [36].

Other studies indicate that, at low RF power and pressures of about 2 mtorr, the dominant reactive species in the oxygen plasma are most probably excited O_2^* molecules and O_2^+ ions in various excited states. In similar plasmas excited in nitrous oxide or ozone, significant amounts of O* species were also observed [37]. Negative ions are also formed in an oxygen plasma, but since they are more massive and are present at low concentrations relative to electrons, they are generally confined to the plasma glow region and play only a minimal role in surface processes.

Ligenza [38] was first to show the possibility of low-temperature oxidation of silicon in a microwave plasma. RF plasmas with various reactor configurations and plasma excitation methods have been studied for plasma oxidation [39–41]. The experiments were performed at substrate temperatures varying between 100 °C and 600 °C. Diverse experimental conditions have been used ranging from DC plasmas where the treated substrate is positively biased to RF sputter etching where the surface is bombarded with energetic ions.

The formation of an oxide on a metal or semiconductor surface immersed in an oxygen plasma at floating potential is called *plasma oxidation*. As the substrate is floating, no current flows through it and the oxide growth proceeds by thermally activated diffusion. The thickness of the oxide thus formed is generally small (less than 100 Å at low temperature) [42]. If the substrate is positively biased above the floating potential, the oxide growth is stimulated and the process is called *plasma anodization*. During plasma anodization, a current flows through the oxide, assisted by the electric field. The oxidation process is accelerated, and thicknesses up to a few micrometers can be obtained. The enhancement of oxide growth as a result of the applied bias indicates that in this case electrically charged particles moving through the oxide contribute to the growth. Plasma anodization is similar to reactive ion etching with the distinction that a solid product, oxide, instead of a volatile one is formed on the surface.

During plasma anodization, plasma-surface interactions involve a large number of charged and neutral species, as illustrated in Fig. 6-4. The positively biased surface immersed in the oxygen plasma is exposed to molecular oxygen and desorbed species and is bombarded by electrons, negative ions, and ions of the oxides of the material sputtered from the substrate [42].

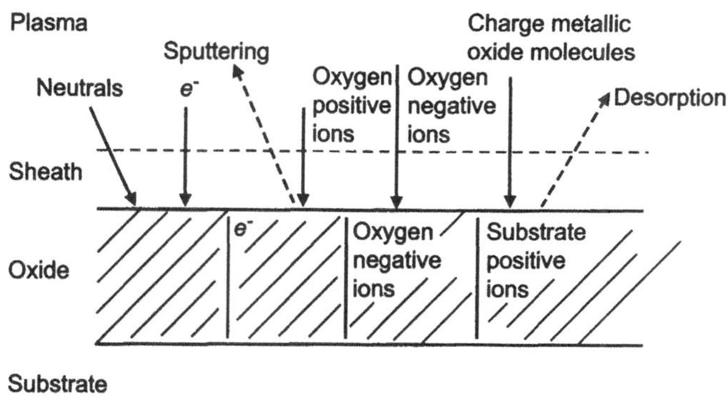

Fig. 6-4 Plasma-surface interactions during plasma anodization (from [42], reprinted with permission from *Plasma Chem. Process*, vol. 1, p. 217, 1981).

It was initially assumed that the growth of the oxide is controlled by the supply of oxygen ions from the plasma; however, there is evidence that the growth rate of the oxide is not limited by the supply and diffusion of negative ions from the gas phase. The growth rate of the oxide has been found to increase with electron temperature. It has also been found that if the electrons are prevented from reaching the treated surface the oxide growth rate decreases strongly. This is an indication that electron bombardment of the surface makes an important contribution to the formation of the negative oxygen ions.

While it was supposed in early studies [38, 43, 44] that the O^- ion is responsible for the oxidation of Si in the plasma, it now appears that oxygen atoms are the most important species responsible for the oxidation and the importance of the negative oxygen ions is not clear [45]. Even when the substrate is positively biased it is not certain whether it undergoes bombardment by negative ions, due to the tendency of the plasma to be always positive relative to a surface exposed to it.

In plasma anodization, when the substrate is positively biased, charge transport takes place in the oxide and the growth of the oxide is controlled by the drift of the negative O^- ions through the film. Electrons and negative ions move inward, while cations of the substrate move outward. The energetic ions penetrate the substrate and create a large electron current. The current efficiency defined as the ratio between the ionic current and the total current is low, of the order of several percent [42]. Under these conditions, the oxidation rate decreases with time, and the oxide reaches a final thickness, which is determined by the negative bias.

Plasma anodization which takes place under synergistic effects of electrons and ions results in enhanced oxidation rates. Plasma anodization rates of 10–100 Å/min, obtained at 300–500 °C, are comparable to those obtainable by thermal oxidation at much higher temperatures of 1100–1200 °C. This is especially important for oxidation of silicon during fabrication of MOS devices, when the processing temperatures decrease continuously.

When the substrate is negatively biased, the negative ions cannot drift to the oxide-metal interface. In this case, the oxide thickness is controlled by diffusion that can be enhanced by bombardment of the surface with positive ions. When ion bombardment is used, an equilibrium can be reached between the oxide growth rate and the rate of removal by sputtering. By adjusting the plasma parameters, such as power, pressure, ratio of oxygen/argon, the thickness of the oxide can be precisely controlled.

The majority of oxygen ions contributing to plasma oxidation are formed by electron assisted surface processes. Ion space charge can control the anodization kinetics. The nature of the oxide surface and electrode contamination can have a determinant role in the exchange of oxygen between plasma and the oxide [42].

Oxidation by direct exposure to the plasma can induce damage in semiconductor substrates. While CV curves measured for the plasma oxides are often the same as for thermal oxides, the density of interface states tends to be one or two orders of magnitude higher in plasma oxides. Breakdown field strengths of 4–8 MV/cm were measured for plasma oxide [38–40].

Studies of oxidation of silicon in an oxygen/argon plasma at 500 °C have indicated that the plasma parameters can have a strong effect on the density of interface states, on the fixed oxide charge, and electron trap density in the plasma oxides [46]. Experimental studies have shown that oxidation of GaAs in an ECR plasma results in the formation of a stoichiometric oxide layer composed mainly of As_2O_5 and Ga_2O_3. A hydrogen plasma could remove that oxide even at room temperature and restore surface order [47].

Two growth regimes have been found in oxidation of silicon and titanium disilicide in a RF plasma at floating condition and temperatures of 400–900°C [48]. The transition between the two regimes occurred at 600 °C. For $TiSi_2$, the nature of the oxide was found to depend on the oxidation temperature: a mixture of TiO_2-SiO_2 was obtained at temperatures below 600 °C, while almost pure SiO_2 was obtained above 600 °C [48]. The oxidation rate was much higher in the low-temperature regime than at the higher one.

High-quality silicon oxide films with excellent interface properties and low-defect density have been obtained by low-rate oxidation in a low-power plasma of oxygen highly diluted in helium. Using the oxidized film as an interfacial layer for PECVD oxide, a high-quality composite layer (plasma oxide/PECVD deposited oxide) with properties similar to thermal oxide was obtained. Such low-temperature oxides were found suitable for semiconductor devices, such as gate oxide in MOS or as base passivation layer in bipolar devices [49]. Prevention of semiconductor damage caused by exposure to the plasma can be achieved by performing the oxidation in the plasma effluent, for example, in a downstream microwave oxygen plasma [50].

During plasma oxidation of silicon in a low-power RF oxygen plasma (at 13.56 MHz and 1 $W.cm^{-2}$), the growth of the oxide film has been observed to take place through an oscillating process [37]. XPS and Fourier transform infrared spectroscopy (FTIR) measurements indicated that the oscillating growth layer was a silicon-rich oxide. Oxide growth in nitrous oxide or ozone plasmas, under the same plasma conditions, followed the normal oxidation-diffusion curve. In the oxygen plasma at the specific conditions, reactive $O_2^{\bullet}$ radicals assisted by ion bombardment with O_2^+ ions absorbed and diffused in the silicon to form a silicon-rich oxide layer [37]. The oscillating oxide growth has been explained to occur as a result of the competition between low-energy oxygen ion assisted plasma oxidation and sputtering [37].

Plasma oxidation has also been used for the preparation of high-T_c superconductors. For example, an ECR oxygen plasma was used to deposit the high-T_c superconductor $HoBa_2Cu_3O_x$ [51]. A high-quality oriented film was deposited on an MgO substrate at 580 °C by ECR plasma assisted reactive evaporation. After evaporation the film was annealed at 620 °C in the same oxygen plasma at a pressure of 10^{-2} Pa. The characterization of the film has shown that the oxygen activity in the ECR plasma is equivalent to that of oxygen without plasma at pressures three orders of magnitude higher.

6.5 SURFACE HARDENING

Although the surface hardening processes are similar to those described previously, especially plasma oxidation, they are discussed separately because different surface properties are modified, and they are used for other applications. Machine parts characterized by high strength and toughness often require an extremely hard surface. The surface hardening processes are used to increase the hardness of surface layers of metals without affecting their bulk properties. These hardening processes include nitriding, carburizing, and boriding.

6.5.1 Nitriding

Nitriding is one of the processes used for surface hardening (case hardening) of parts requiring increased wear resistance and increased fatigue life. Nitriding can also improve the corrosion resistance of the treated metals. The hardness achieved by nitriding is maintained even at high temperatures. A case of high hardness with depth up to 75 μm can be obtained without any additional heat treatment, thereby avoiding distortion of the parts. A wide variety of metals and alloys can be nitrided: cast iron, low-alloy steels containing aluminum and chromium, tool steels, stainless steels, titanium and its alloys, and so on.

Conventional thermal nitriding is performed in partially dissociated ammonia at temperatures between 480 °C and 650 °C. The treatment time is lengthy and can take up to 50 hours to obtain the desired case depth. Full hardness is achieved directly after nitriding, and no supplementary treatment is required.

Plasma nitriding, or as it is often called *Ionitriding*[1], is a surface modification process similar to plasma oxidation. However, in contrast to oxygen, nitrogen does not form a negative ion in the plasma or near the treated surface. Therefore, there is no significant contribution to the nitriding process from field assisted motion of negative ions toward the nitride-metal interface. Plasma nitriding is best accomplished as a cathodic process with concomitant bombardment by energetic positive ions.

In the ionitriding process the parts are inserted into a chamber that can be kept at a vacuum between 0.1 and 10 torr. The pressures, generally between 2 and 5 torr, can be adjusted to optimize penetration into the cavities of the processed parts. A DC glow discharge is produced in the chamber where the parts to be nitrided serve as the cathode of the discharge, while the walls of the chamber serve as the anode. DC voltages between 100 and 1500 V are used. The plasma is sustained in the anomalous glow discharge region of the discharge characteristic (see Fig. 2-2 in Chapter 2).

The parts are heated by bombardment with highly energetic positive ions, and the temperatures reached (400–600 °C) depend on the discharge power. Both atomic and molecular nitrogen contribute to the nitriding process. At low tem-

[1] Ionitriding is a trademark of Klockner Ionon GmbH.

peratures, when the diffusion rates are low, the growth of the nitride film is determined by the enhanced diffusion caused by energetic ion bombardment. In rapid nitride growth, for which high current and power densities are desirable, the abnormal region of the glow discharge is preferred. Typical current densities, J, and voltages, U, used in plasma nitriding are

$$J = 0.1\text{–}10 \text{ mA/cm}^2$$

$$U = 500\text{–}1000 \text{ V}$$

The large number of parameters, which can be controlled separately, make possible to produce various structures of the nitride layer not obtainable in conventionally nitrided metals.

Ionitriding is done either in pure nitrogen or in mixtures of nitrogen with hydrogen. Vacuum furnaces with working volumes ranging from 0.1 to 8 m^3 are used for plasma nitriding. The DC power supplies can reach an output of up to several hundred kilowatts, corresponding to currents of several hundreds amperes at 1000 V [52]. Parts that are plasma nitrided range from 0.7 mm diameter ballpoint pen balls to units weighing several tons.

In addition to absorbing neutral gas species, the cathode undergoes bombardment with energetic ions. It is believed that nitrogen-hydrogen molecular ions (NH^-) play a vital role in the ionitriding process. Replacement of hydrogen with argon decreases the development of the hardened case, indicating that the sputtering of the surface (more efficient with argon than with hydrogen) has only a secondary effect in plasma nitriding. A mixture of nitrogen with hydrogen (25 to 50% N_2 for steels) has been found to give the best results in ionitriding. Replacement of hydrogen by other gases results in reduced rate of case development.

In the DC plasma, most of the applied voltage drops within only a few millimeters in front of the cathode. Thus most of the ionization occurs in the vicinity of the cathode, in the cathode fall region, and the surface treatment is practically independent from the distance between the treated part and the walls of the reactor. The cathodic glow is limited to a narrow region around the treated parts and follows their contour, including notches and bores, indicating a uniform ion bombardment of the parts that results in a uniform hardening.

The plasma nitriding is a result of several processes that occur in the reactor [52]:

1. *Sputtering.* The energetic ions that hit the surface cause the sputtering of atoms from the surface. In addition to the sputtering of components of the metal alloy, surface impurities are also sputtered away, leading to a cleaning of the surface. Removal of oxygen results in to a complete surface depassivation, even of high-chromium or stainless steels.
2. *Heating.* The energy balance of an ion hitting the surface of the metal and causing the sputtering of a surface atom can be written as

$$W_{ion} = W_f + W_{kin} + Q \tag{6.1}$$

where W_{ion} = kinetic energy of incident ion
W_f = work used to sputter an atom from the surface
W_{kin} = the kinetic energy of the sputtered particle
Q = remaining energy converting into heat

It becomes clear from Eq. (6.1) that the ions hitting and penetrating the surface layer impart the fraction Q of their energy to the workpiece and heat it. The temperature reached is a function of the ion energy (applied voltage) and ion density (current).

3. *Condensation.* Only a small fraction of the nitriding is a result of direct incorporation of nitrogen atoms into the treated surface. During plasma nitriding of steel, for example, iron atoms sputtered from the surface react with the highly active nitrogen atoms in the plasma to form iron nitride (FeN), which redeposits on the surface. This nitride is unstable and decomposes into lower nitrides Fe_2N, Fe_3N, Fe_4N, as illustrated in Fig. 6-5.

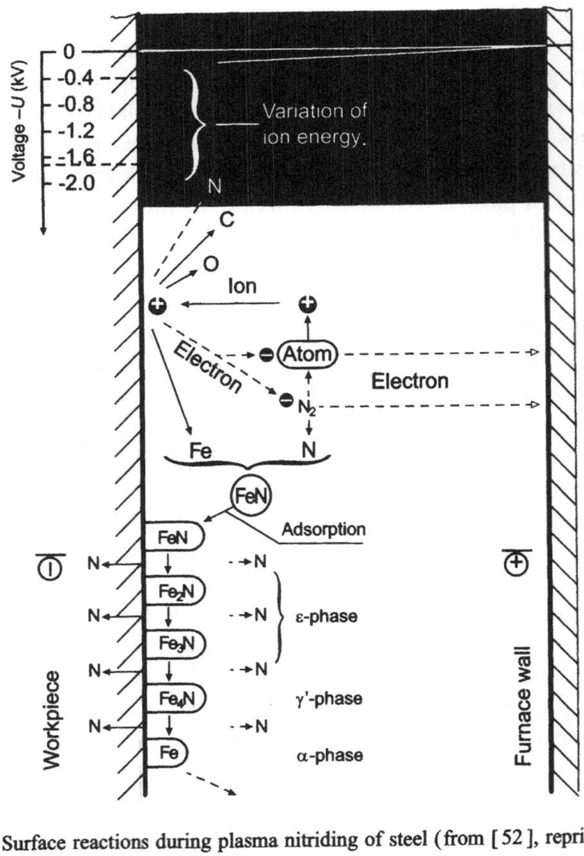

Fig. 6-5 Surface reactions during plasma nitriding of steel (from [52], reprinted with permission.).

The nitrogen, liberated by the decomposition of FeN, diffuses into the workpiece. The abundant supply of nitrogen creates a steep concentration gradient already at the beginning of the process. As a result of the diffusion of the nitrogen into the metal, the iron saturates with nitrogen and more iron nitrides are formed. Because a high nitrogen concentration forms at the surface of the metal at the start of the process, the ionitriding is much faster than conventional nitriding. Moreover, the composition of the nitrided layer can be controlled by the process parameters of the plasma, a very significant advantage because the properties of the nitrided layer depend strongly on its composition.

A study of plasma nitriding of the Ti-6Al-4V alloy indicated that the nitriding occurs along several paths [53]:

1. Formation of TiN as a result of the reaction between active nitrogen species from the plasma and titanium.
2. Depletion of the alloy of aluminum and vanadium and formation of a compound layer of ε-(Ti_2N) + δ-TiN crystallographic phases.
3. Solution of nitrogen in the a phase and formation of the a'-(Ti,N) phase.

The addition of hydrogen to the nitriding plasma enhances the second step, while the addition of argon reduces the solubility of nitrogen in titanium and promotes the formation of the ε phase.

The variety of properties of the nitrided metal that can be achieved by modifying the plasma parameters is illustrated for the nitriding of Ti-6Al-4V alloy [53]. It was found that the nitrided layer at the surface of the Ti-6Al-4V alloy is composed of an external compound layer, followed by a diffusion layer. The compound layer consists of cubic δ-TiN and tetragonal ε-Ti_2N phases. The diffusion layer is composed of the ε-Ti_2N phase and a solid solution of nitrogen in titanium a'-(Ti,N). The relative concentration of the different phases and their crystallographic orientation, as well as the mechanical properties of the treated surface, have been found to be strongly dependent on the plasma parameters, such as composition of discharge gas, substrate bias, frequency of the discharge, and position in a tubular reactor. This dependence is illustrated in Fig. 6-6.

The results in Fig. 6-6 are presented for nitriding at two positions in an inductively coupled RF plasma: (a) G — at the center of the RF coil; (b) F — downstream from the center of the plasma. The composition of the nitriding gas mixtures are shown at the top of the plots. The values on the abscissa are as follows: (i) microhardness, (ii) erosion reduction, and (iii) number of cycles to fatigue failure.

Significantly higher surface hardness is obtained for the Ti-6Al-4V alloy by nitriding at position G, versus position F, with the maximum hardness obtained by nitriding with a gas mixture containing 20% hydrogen in nitrogen. The hardening is most strongly correlated to the amount of the δ phase and its stoichiometry. Resistance to erosion was improved by nitriding the alloy, and the improvement was directly related to the surface microhardness and thus the content of δ phase

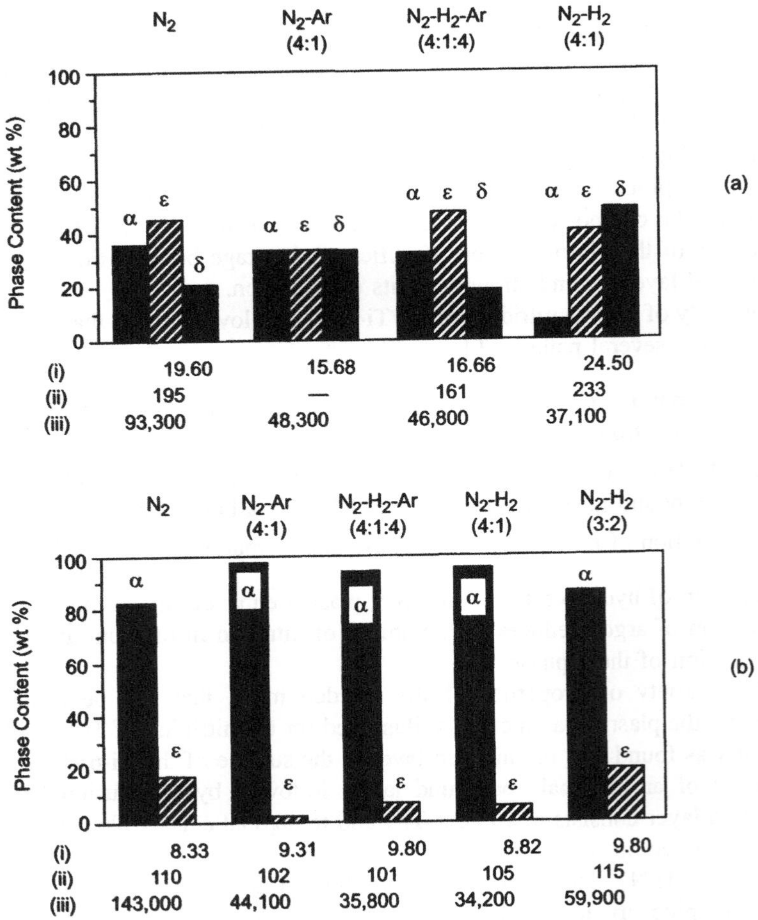

Fig. 6-6 Mechanical properties and phase composition of nitrided Ti-6Al-4V as a function of nitriding parameters (from [53], reprinted with permission from *Surf. Coat. Technol.*, vol. 43/44, p. 745, 1990): (a) nitriding at center of plasma; (b) nitriding downstream from the center. The abscissa values are (i) microhardness (kgf.mm^{-2}); (ii) erosion reduction (%); (iii) number of cycles to fatigue failure.

in the nitrided layer. Significant improvement of the fatigue life of the alloy could also be obtained under certain nitriding conditions. By choosing the right combination of plasma parameters, it is possible to tailor the surface properties of the plasma nitrided alloy according to the desired end product.

The ionitriding presents several advantages over conventional nitriding, among them:

- Versatility of the process in terms of control of layer properties and of workpiece material

- Wide range of treatment temperatures, between 300 °C and 600 °C
- Control of the structure and properties of the nitrided layer
- Reduced treatment time, 3 to 30 hours as compared to 50 hours in conventional nitriding.

The consequence of plasma nitriding performed at lower temperatures than conventional nitriding is generally higher-dimensional stability of the processed items which makes this method suitable for hardening of precision machine parts.

6.5.2 Carburizing

Carburizing is generally obtained by increasing the carbon content of the surface layers of low-carbon or alloy steels followed by heat treatment to obtain surface hardness. Plasma carburizing is similar in principle to ionitriding. The plasma primarily assists in the mass transfer of carbon atoms to the surface of the treated sample. Carburizing of steel has been performed in a DC discharge of a hydrocarbon gas at 1–20 torr, with additional heating of the parts at 1050 °C [54, 55]. Levels of carbon in excess of 1% were observed after 5 minutes of plasma carburizing.

It was found that plasma enhanced the carburizing rate, and it has been suggested that the high infusion rate of carbon into the surface of the steel, and not the high diffusion rate, is responsible for the high carburizing rate. The increase of carbon concentration in the surface layer was observed to occur almost instantaneously during plasma carburizing [56]. The plasma treatment is performed for a short time, but heating is performed for longer periods of time to permit the diffusion of carbon in steel.

Unlike other plasma hardening processes, plasma carburizing of steels is not performed at reduced treatment temperature, because high temperatures are required for diffusion of carbon in the austenitic phase. The plasma carburized parts have to be generally ground to final shape after the treatment, a step not commonly required for nitrided parts. The need for grinding is a consequence of the fact that carburizing, even when obtained by plasma processing, needs much higher temperatures than nitriding [57].

The advantages of plasma carburizing are in the obtained uniformity of the carburized layer and reduction of treatment time. Uniformity is attained since the plasma follows the profile of the treated parts. The plasma also enhances both the mass transfer of the carburizing gas to the surface and the reactions occurring at the surface of the workpiece.

Carbonitriding is another plasma hardening process well accepted in commercial production. Plasma carbonitriding is a combination of carburizing and nitriding of steel and is performed in gas mixtures of nitrogen with methane, the latter providing the carbon atoms [58].

6.5.3 Boriding

Boriding is a surface hardening method for improving wear resistance of steels and sometimes their corrosion and oxidation resistance. The surface hardness is achieved in boriding by the formation of borides during the treatment process. Conventional boriding is generally performed at temperatures above 900 °C. It was shown that boriding can also be performed by plasma treatment in a mixture of diborane with argon [59]. The surface hardening was achieved at a treatment temperature of only 600 °C. The boriding in the plasma was found to proceed through the deposition of a boron layer on the surface of the sample, by the dissociation of the diborane in the plasma. The boron diffused into the steel to form FeB and Fe_2B boride phases, and surface hardness increased from 2.65 GPa to 9.8 GPa after the plasma boriding [59].

6.6 QUESTIONS

1. What kind of plasma species can be used for surface modification of polymers and what type of modification can be produced by each?
2. What plasma processes can be used for cleaning of solid surfaces? Which gases can be used with these plasmas?
3. Describe the plasma ashing process. What are its advantages compared to other ashing methods?
4. Discuss the differences between plasma and thermal oxidation? What are the advantages of plasma oxidation?
5. Which are the differences between plasma oxidation and plasma anodization?
6. a. What is the nitriding process and what are its applications?
   b. Compare plasma nitriding with the conventional nitriding techniques.
7. What is the mechanism of plasma nitriding? What advantages result from this mechanism?

6.7 REFERENCES

[1] Coburn, J. W., and H. F. Winters, *J. Vac. Sci. Technol.*, 16: 371 (1979).

[2] Beauchamp, J. L., and S. E. Buttrill, *J. Chem. Phys.*, 48: 1783 (1968).

[3] Siperko, L. M., and R. H. Thomas, *J. Adhesion Sci. Technol.*, 3: 157 (1989).

[4] Hoffman, A. S., In *Plasma Polymerization and Plasma Interactions with Polymeric Materials*, ed. H. K. Yasuda, p. 341. New York: John Wiley & Sons, 1990.

[5] Yasuda, H. K., ed., *Plasma Polymerization and Plasma Interactions with Polymeric Materials*. New York: John Wiley & Sons, 1990.

[6] Cormia, R., *Res. Dev. Mag.*, 32(7): 60 (1990).

[7] Yasuda, H. K., *Plasma Polymerization*. New York: Academic Press, 1985.

[8] Kramer, P. W., Y. Yeh, and H. Yasuda, *J. Membr. Sci.*, 46: 1 (1989).

[9] Inagaki, N., S. Tasaka, J. Ohkubo, and H. Kawai, In *Plasma Polymerization and Plasma Interactions with Polymeric Materials*, ed. H. K. Yasuda, p. 399. New York: John Wiley & Sons, 1990.

[10] Momose, Y., T. Ohaku, H. Chuma, S. Okazaki, T. Saruta, M. Masui, and M. Takeuchi, In *Plasma Polymerization and Plasma Interactions with Polymeric Materials*, ed. H. K. Yasuda, p. 153. New York: John Wiley & Sons, 1990.

[11] Kusabiraki, M., In *Plasma Polymerization and Plasma Interactions with Polymeric Materials*, ed. H. K. Yasuda, p. 473. New York: John Wiley & Sons, 1990.

[12] Hollahan, J. R., B. B. Stafford, R. D. Falb, and S. T. Payne, *J. Appl. Polymer Sci.*, 13: 807 (1969).

[13] Egitto, F. D., V. Vukanovic, and G. N. Taylor, In *Deposition, Treatment and Etching of Polymers*, ed. R. d'Agostino, p. 321. New York: Academic Press, 1990.

[14] Molis, S. E., D. G. Kim, S. P. Kowalczyk, and J. Kim, *Mat. Res. Soc. Symp. Proc.*, 203: 53 (1991).

[15] Morosoff, N., In *Deposition, Treatment and Etching of Polymers*, ed. R. d'Agostino, p. 1. New York: Academic Press, 1990.

[16] Bradley, A., *Chem. Technol.*, 3: 507 (1975).

[17] Meichsner, J., and H. U. Poll, *Acta Polym.*, 22: 203 (1981).

[18] Matsuda, Y., and H. Yasuda, *Thin Solid Films*, 118: 211 (1984).

[19] Ruzyllo, J., *Solid State Technol.*, 33(3): S1 (1990).

[20] Sugano, T., ed., *Applications of Plasma Processes to VLSI Technology*. New York: John Wiley & Sons, 1985.

[21] Yew, T. R., and R. Reif, *J. Appl. Phys.*, 68: 4681 (1990).

[22] Yew, T., and R. Reif, *Mat. Res. Soc. Proc.*, 202: 401 (1991).

[23] Shibata, T., Y. Nanishi, and M. Fujimoto, *Jpn. J. Appl. Phys. Lett.*, 29: 1181 (1990).

[24] Ruzyllo, J., D. C. Frystak, and R. A. Bowling, *Proc. IEDM*, 90: 409 (1990).

[25] Yoon, E., and R. Reif, *Mat. Res. Soc. Proc.*, 202: 37 (1991).

[26] Mukkavilli, S., R. W. Pasco, M. S. Farooq, and M. J. Griffin, In *Proc. 40th Electronic Components & Technology Conf.*, May 20–23, 1990, Las Vegas, NV (IEEE, 1990), p. 737.

[27] Hollahan, J. R., In *Techniques and Applications of Plasma Chemistry*, eds. John R. Hollahan and Alexis T. Bell, p. 229. New York: John Wiley & Sons, 1974.

[28] Thomas, R. S., In *Techniques and Applications of Plasma Chemistry*, eds. John R. Hollahan and Alexis T. Bell, p. 255. New York: John Wiley & Sons, 1974.

[29] Cecchi, J. L., In *Handbook of Plasma Processing Technology*, eds. Stephen M. Rossnagel, Jerome J. Cuomo, and William D. Westwood, p. 14. Park Ridge, NY: Noyes Publications, 1990.

[30] Hartney, M. A., D. W. Hess, and D. S. Soane, *J. Vac. Sci. Technol.*, B7: 1 (1989).

[31] Flamm, D. L., *Plasma Chem. Plasma Process.*, 1: 37 (1981).

[32] Bersin, R. L., *Solid State Technol.*, 13: 391 (1970).

[33] Spencer, J. E., R. A. Borel, and A. Hoff, *J. Electrochem. Soc.*, 133: 1922 (1986).

[34] Bell, A. T., In *Techniques and Applications of Plasma Chemistry*, eds. John R. Hollahan and Alexis T. Bell, p. 1. New York: John Wiley & Sons, 1974.

[35] Thomson, J. B., *Proc. R. Soc. London Ser. A*, 262: 503 (1961).

[36] Bell, A. T., and K. Kwong, *AIChE J.*, 18: 990 (1972).

[37] Nguyen, S., T. Nguyen, D. Dobuzinsky, R. Gleason, and M. Gibson, *Mat. Res. Soc. Proc.*, 223: 229 (1991).

[38] Ligenza, J. R., *J. Appl. Phys.*, 36: 2703 (1965).

[39] Pulfrey, D. L., F. G. M. Hathorn, and L. Young, *J. Electrochem Soc.*, 120: 1529 (1973).

[40] Ray, A. K., and A. J. Reisman, *J. Electrochem Soc.*, 128: 2466 (1981).

[41] Ho, V. Q., and T. Sugano, *IEEE Trans. Electron. Devices.*, ED-27: 1436 (1980).

[42] Gourier, S., and M. Bacal, *Plasma Chem. Plasma Process.*, 1: 217 (1981).

[43] Padushek, P., *Technical Proceedings SEMICON EUROPA*, Zurich, Switzerland, 10-12 March, 1987, p. 107.

[44] Perriere, J., J. Siejka, N. Rmili, A. Laurent, A. Straboni, and B. Vuillermoz, *J. Appl. Phys.*, 59: 2752 (1986).

[45] Coburn, J. W., In *Diamond and Diamond-like Films and Coatings, NATO-ASI Series B: Physics*, eds. R. E. Clausing, L. L. Horton, J. C. Angus, and P. Koidl, Vol. 266, p. 73. New York: Plenum Press, 1991.

[46] Chokski, A. J., R. Lal, and A. N. Chandorkar, *Solid State Electron.*, 34: 765 (1991).

[47] Lu, T. M., P. Li, E. J. Rymaszewski, H. J. Greub, and J. F. McDonald, *Proceedings of the Technical Program — National Electronic Packaging and Production Conference*, 3: 1833-1842 (1991).

[48] Perez, R. C., J. Perriere, J. P. Enard, A. Straboni, B. Vuillermoz, A. Climent, and J. M. M. Duart, *J. Appl. Phys.*, 69: 1407 (1991).

[49] Bright, A. A., J. Batey, and E. Tierney, *Appl. Phys. Lett.*, 58: 619 (1991).

[50] Salbert, G. T., D. K. Reinhard, and J. Asmussen, *1989 IEEE Int. Conf. Plasma Sci.*, 155 (1989).

[51] Tsukamoto, A., M. Hiratani, and S. Akamatsu, *Physica C*, 181: 369 (1991).

[52] Edenhofer, B., *Heat Treat. Met.*, 23 (1974).

[53] Grill, A., A. Raveh, and R. Avni, *Surf. Coat. Technol.*, 43/44: 745 (1990).

[54] Grube, W. I., and J. G. Gay, *Metal. Trans.*, A9: 1421 (1978).

[55] Staines, A. M., and T. Bell, *Thin Solid Films*, 76: 201 (1981).

[56] Jacobs, M. H., and W. Metalsky, In *Proc. 6th ISPC*, eds. M. I. Boulos and R. J. Munz, p. 737. Montreal: McGill Univ., 1983.

[57] Manory, R., and A. Grill, *Plat. Surf. Finish.*, 73: 68 (1986).

[58] Spalvins, T., *Thin Solid Films*, 108: 157 (1983).

[59] Grill, A., I. Finberg, R. Avni, T. Spalvins, and D. H. Buckley, *Mat. Sci. Eng.*, 78: 201 (1986).

7

Deposition of Coatings by PECVD

7.1 DEPOSITION OF ORGANIC FILMS

7.1.1 Plasma Polymerization

Plasma polymerization is essentially a PECVD process resulting in the deposition of an organic polymer. It refers to the deposition of polymer films through plasma dissociation and excitation of an organic monomer gas and subsequent deposition and polymerization of the excited species on the surface of a substrate. Plasma polymerization is used to deposit films with thicknesses from several tens to several thousands of angstroms. The deposited films are called *plasma polymers* and are generally chemically and physically different from conventional polymers [1]. Hydrocarbons, fluorocarbons, and organic compounds containing nitrogen, oxygen, and silicon have been used as starting materials for plasma polymerization. Detailed discussion of plasma polymerization processes can be found in various reviews and books, such as references [2–6]. Only the basic phenomena of plasma polymerization and selected examples will be discussed in this section.

Plasma polymerization is a process that results in the preparation of a new type of material and is not a kind of polymerization. The materials formed by plasma polymerization are very different from conventional polymers and are also different from most inorganic materials. Plasma polymerization is a strongly system-dependent process. Conventional polymerization is based on molecular processes during which rearrangements of the atoms within the monomer seldom occurs. In contrast to such molecular processes, polymer formation in a plasma is an atomic process. Polymers formed by plasma polymerization are in most cases highly branched and highly cross-linked. Free radicals formed on the surface of polymers and other solid materials exposed to a plasma can be used to initiate graft polymerization.

Plasma polymerization is characterized by several features [1, 5]:

1. Plasma polymers have no discernible repeating units as do conventional polymers.
2. The properties of the plasma polymer are not determined by the used monomer but by the plasma parameters; for example, there is not a product that can be identified as the plasma polymer of ethylene, because a variety of products can be obtained from an ethylene plasma; plasma polymers of ethylene are not polyethylene.
3. The monomer used for plasma polymerization does not have to contain a functional group, such as a double bond, for the plasma polymerization to take place.

Plasma polymerization can be used to prepare a variety of products, which include oils, powders, soluble polymers, and insoluble films.

From the point of view of plasma deposition of polymers, plasmas can be divided, according to the precursors, into nonpolymer-forming plasmas and polymer-forming plasmas. In the first type of plasmas, the feed gas, such as nitrogen, hydrogen, or ammonia, is nonpolymerizable. In the second type the feed gas, or monomer, is polymerizable in the plasma in the sense that itself, or parts of it, can be used as the building block of macromolecules. Plasmas of methane or styrene are considered to be polymer-forming plasmas, although methane cannot be polymerized by conventional methods [1]. Almost all organic compounds that have sufficient vapor pressure to be fed into a plasma are polymerizable. Monomers used for plasma polymerization include hydrocarbons, fluorocarbons, sulfides, siloxanes, silanes, and so on.

As described in Section 3.4, plasma polymerization takes place through several reactions steps, that is, initiation, propagation, termination, and reinitiation [4]. In the initiation stage, free radicals and atoms are produced either by collision of energetic electrons and ions with monomer molecules or by dissociation of monomers adsorbed on the surface of the sample due to impact energetic ions. Both monomer molecules and radicals are adsorbed on the surface exposed to the plasma. The propagation of the reactions, that is, the formation of the polymeric chain, can take place both in the gas phase and on the deposited polymer film. In the gas phase, propagation involves the addition of a radical atom to another radical or molecule. At the surface of the polymer film propagation occurs through interactions of surface free radicals with either gas phase or adsorbed monomers. Termination also takes place either in the gas phase or at the polymer surface by processes similar to the propagation step but ending either with the final product or a closed polymer chain.

While in conventional polymerization the termination step interrupts the process, in plasma polymerization neutral products formed in the termination step can undergo reinitiation and propagation reactions. In the reinitiation processes, chain fragments can be reconverted into radicals by collision with electrons in the gas phase or by impact of energetic particles or photon absorption on the surface of the polymer film.

The described homogeneous polymerization in the plasma phase and heterogeneous polymerization at the surface of the growing film can be summarized by the reaction scheme proposed originally by Poll et al. [7] and illustrated in Fig. 7-1. As shown in this figure, the gaseous precursor monomers can polymerize in the plasma and deposit as a film (reaction path 1, characterized by a rate k_1), be converted in the plasma into reactive products (path 2), or be converted in nonreactive products (path 4). The reactive products can convert into depositing

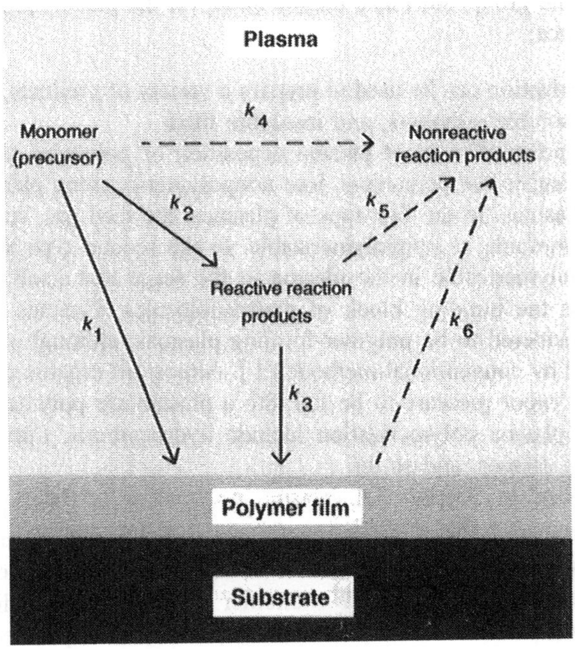

Fig. 7-1 Diagram of Plasma Polymerization (after [7], with kind permission from Pergamon Press Ltd, Headington Hill Hall, Oxford 0X3 0BW, UK). Solid line, polymer deposition path; broken line, volatile product formation path.

polymer film (path 3) or can be converted to nonreactive products (path 5). Degradation of the formed polymer film can occur to form nonreactive products (path 6).

Path 1, by which the monomer is directly polymerized into the growing film, is also called *plasma induced polymerization*. This is essentially a conventional molecular polymerization process, triggered in this case by reactive plasma species. Plasma induced polymerization can take place only if the original monomer contains polymerizable functional groups, such as double, triple, or cyclic bonds [2].

Deposition of a polymer film through paths 2 and 3 is *plasma polymerization*. In plasma polymerization the intermediate reactive products can be ions, excited molecules, and free radicals not necessarily preserving the original monomer [2]. The monomer does not need to have polymerizable groups to undergo plasma polymerization. Compounds such as tetramethylsilane (TMS) or tetramethyltin (TMT) are nonfunctional and therefore do not polymerize by conventional polymerization. In a plasma, the sylmethyl and tinmethyl units dissociated by electron impact become functional groups and can polymerize. Ethane and methane can also be polymerized in a plasma. Using unsaturated fluorocarbon gases, such as tetrafluoroethylene or hexafluoropropene, Teflon®-like films can be deposited. A polymer similar to polystyrene, but not identical to conventional polystyrene, can be formed in a plasma from a styrene monomer.

In the following paragraphs *plasma polymerization* will, however, refer to *all polymerization processes* taking place in a cold plasma.

7.1.2 Polymerization Mechanisms

Conventional polymerization reactions can generally be classified in two groups, according to the mechanism of polymer growth [5]:

1. Step-growth polymerization
2. Chain-growth polymerization

In step-growth polymerization, a polymer is formed by the stepwise repetition of the same reaction over and over again. The product of the reaction retains functional groups that can react further the same way as the original material and the reaction at each step is identical to the first reaction. The growth rate is slow, the molecular weight of the polymer depends on the reaction time, and the components of the reaction mixture change with time [5].

In chain-growth polymerization a long-chain molecule is formed by a series of consecutive steps that are completed in a very short time, generally a fraction of a second. The products of the reaction are only final polymers; intermediate-size molecules cannot be isolated as in the case of step-growth polymerization. The chain-growth reaction is very fast, the molecular weight of the product does not depend on the reaction time, and the reaction mixture is composed only by the original monomer and final polymer.

In plasma polymerization chain-growth polymerization plays an insignificant role, and according to Yasuda [5], plasma polymerization takes places through the *rapid step-growth polymerization* (*RSGP*) mechanism, also called *activation-growth model*, which can be described by the equation

$$(M_m^* + M_n^* \rightarrow M_{m+n}) \times N \tag{7.1}$$

where N = the number of repetitions of similar reactions
$M^{\bullet}$ = the reactive chain-carrying species produced in the plasma from monomer M.

If the reactive species $M^{\bullet}$ is monofunctional, such as a free radical $R^{\bullet}$, the reaction $M_n^{\bullet} + M_n^{\bullet}$ is essentially a termination process and does not contribute to growth continuation without additional elementary steps. Such a step can be a reactivation reaction that can cause the formation of a new functional, which in turn allows the continuation of the chain growth.

Plasma polymerization through the rapid step-growth mechanism is illustrated in Fig. 7-2. In the figure, M_x describes a neutral species that can be the original precursor monomer or a dissociation product including atoms. Bifunctional activated species are marked in Fig. 7-2 by $^{\bullet}M^{\bullet}$. The subscripts indicate differences in the sizes of involved species.

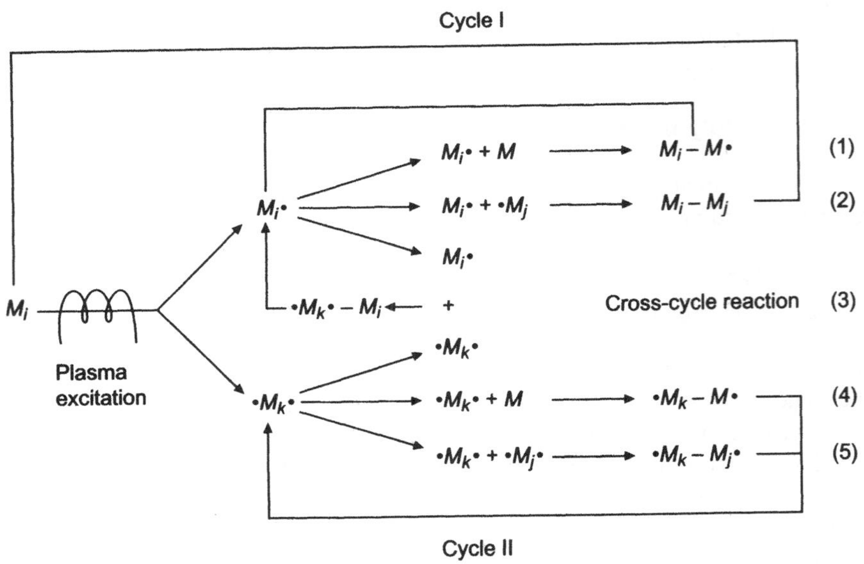

Fig. 7-2 Schematic representation of step growth mechanism of plasma polymerization (from [5], reprinted with permission from *Plasma Polymerization*, Fig. 6.1, 1985).

According to Fig. 7-2 the RSGP can proceed through two parallel cycles. Cycle I takes place through repeated activation of the reaction products from monofunctional activated species; cycle II occurs through difunctional or multiactivated species. Plasma diagnostics showed that plasmas of organic compounds contain large amounts of bifunctional species such as $\cdot CH \cdot$ or $\cdot CF \cdot$. Cycle I and cycle II may therefore play equivalent roles in plasma polymerization. In both cases the growth takes place through the RSGP mechanism. Reactions (2-1) and (2-4) are essentially the same as the first step of propagation by the addition mechanism. Reaction (2-2) describes a termination by recombination mechanism. Reaction (2-3) is similar to (2-2), but one reactant is bifunctional. Reaction (2-4) is similar to (2-1), except that a reactant is bifunctional. Reaction (2-5) is a combination of bifunctional intermediates [5].

Ionic species can contribute to the polymerization through ion-molecule reactions [8–10], or through the interactions between positive and negative ions. Such a contribution of ionic species would be possible only in a plasma that contains a considerable quantity of negative ions (e.g., plasmas of halogens and/or oxygen). However, it appears that free radicals are the dominant species controlling the polymerization by radical-molecule or radical-radical reactions [11–13]. The ionization energy of organic molecules is generally larger than 10 eV, which is much larger than the bond energies of primary bonds [5]. Therefore, low-energy atoms that have insufficient energy to ionize the precursors can nevertheless break the molecular bonds and create excited species that trigger chemical reactions. At pressure of about 1 torr the free radical concentration in a cold plasma is up to 10^5 times higher than the ion concentration. Therefore, radical-molecule reactions occur generally at faster rates than do ion-molecule reactions.

Among the possible neutral activated species that can participate in the RSGP mechanism described in Fig. 7-2, the free radical is the most probable and important species. The RSGP mechanism, based on free radicals as the activated species, does not require any special precursor structure to obtain a plasma polymer.

The role of the radicals in plasma polymerization was illustrated by Kobayashi et al. [11], who showed that addition of haloforms to hydrocarbon monomers accelerates significantly the polymerization rate and reduces the H/C ratio in the polymer. This was a result of the increase of radical concentration in the plasma caused by abstraction of hydrogen by the halogens (through a disproportionation type reaction such as Eq. (3.35) in Chapter 3).

Residual free radicals involved in the RSGP mechanism remain in the polymer as trapped free radicals [5]. Monomers containing triple bonds yield plasma polymers with the largest number of trapped free radicals, followed by monomers containing double bonds, which yield polymers with a moderate number of trapped free radicals. Polymer formation proceeding via cycle II in Fig. 7-2 can lead to formation of polymers with unquenched free radicals. A high concentration of free radicals has indeed been observed at the surface of polymers exposed to a plasma [12], and these radicals are capable of reacting with unsaturated monomers.

An important and unique feature of plasma polymerization is the incorporation of gases that do not by themselves form polymers or solid deposits in a plasma. This incorporation of gases is a *plasma copolymerization* process and not trapping of gas molecules in plasma polymers. An example is the incorporation of nitrogen [5].

The complexity of the polymerization paths in the plasma is illustrated by the following examples. Studies of polymerization of propylene or propylene with argon in a microwave plasma have shown that the polymerization takes place via three main reaction chains, namely, the propylene, acetylene, and ethylene chains [14]. Plasma polymerization of tetrafluoroethylene (TFE), was studied in both capacitive and inductive discharge at 13.56 MHz. In the capacitive plasma, at pressures of 500 mtorr, deposition occurred on the electrodes. At low pressures, below 100 mtorr, deposition occurred on a substrate inserted midway between the electrodes. In the inductively coupled discharge a major fraction of the polymer deposited on the walls of the reactor. The formed polymer was found to be similar in both capacitive and inductive RF discharges, but was found to be completely different from conventional poly-(TFE) [15].

Attempts have been made to limit the degree of dissociation during deposition, so that the functional group of the monomer, such as oxygen- or nitrogen-containing groups, is preserved and incorporated in the film. Among the methods investigated for reduced monomer fragmentation are pulsed RF discharges and remote plasma deposition. In the latter it is possible to excite an inert gas plasma in the energetic region of the plasma reactor and to introduce the monomer in the downstream plasma. Minimization of precursor fragmentation can also be achieved by lowering the temperature of the substrate during deposition. Sufficiently low-substrate temperature causes increased precursor adsorption or condensation to occur and consequently increases the fraction of unfragmented precursor incorporated in the film.

It has also been found that functional groups of the precursor can be incorporated to a higher extent in the plasma polymer when deposition is done on cooled substrates [16]. Plasma polymerization of 2-hydroxyethyl methacrylate (HEMA), ring opening ethylene oxide, and nonpolymerizable acetone precursors have been examined as a function of substrate temperature [16]. It has been found that, by reducing the substrate temperature during plasma polymerization, the retention of precursor moieties in the final product can be increased. Both HEMA and ethylene were incorporated into the film deposited at − 80 °C with little fragmentation. Even plasma polymerized acetone deposited at low temperature (− 75 °C) showed increased retention of oxygen.

Methods that reduce monomer fragmentation generally result in reduced deposition rate of the plasma polymer.

7.1.3 Plasma Polymerization Parameters

The most distinguished characteristic of plasma polymerization is that the chemical structure of plasma polymer obtained from same monomer is not unique but is largely controlled by reaction parameters. The plasma parameter that appears to control the deposition of polymer film under constant conditions of pressure, plasma excitation frequency, and monomer is the power (P) to flow-rate (Q) ratio, (P/Q)[1] [15]. The parameter in fact reflects the effect of n_e, which

[1] While the W/F notation is generally used in the literature, the P/Q notation will be adopted here to make it consistent with the symbols used throughout the book.

increases with increasing power, and residence time which in turn increases with Q^{-1}.

A modified plasma parameter, which also takes into consideration the molecular weight of the monomer, *M*, is $P/(QM)$ [5, 17]. This parameter indicates the energy input per unit mass of monomer. For homologous monomers in similar reactors, the behavior of the process parameters as a function of $P/(QM)$ is generally not affected by the molecular weight of the monomer. According to the RSGP model, the chemical structure of plasma polymer is controlled by the fragmentation pattern of the monomer in the plasma. Because the dissociation of a monomer into reactive fragments in the plasma should be related to the specific energy received per unit molecule, $P/(QM)$, the control of chemical structure by this parameter is well rationalized [18]. It was indeed found that the chemical structure of a plasma polymer obtained at a specific value of $P/(QM)$ was unique for different choices of individual *P* and *QM* [18].

The range of structure control by $P/(QM)$ extends over two completely different types of materials that are distinguished from each other in their basic chemical structures:

- Polymers, which are composed of linear covalently bonded monomers
- Inorganic materials, which are composed of three-dimensional tight covalent networks.

At relatively low values of $P/(QM)$ a plasma polymer is obtained whose structure and properties are very similar to those of a conventional polymer. With increasing $P/(QM)$, the chemical structure changes gradually to the more dense covalent network with less organic groups, such as CH_3 or CH_2, until it is almost completely composed of tetrahedrally bonded inorganic structures at extremely high $P/(QM)$ values. Most plasma polymers, which are distinguished from conventional polymers, consist of a structure that lies somewhere between the polymeric and inorganic materials and are in the form of a highly cross-linked and highly branched three-dimensional network [5].

The experimentally observed dependence of the chemical structure on $P/(QM)$ cannot be explained by the chain-growth polymerization mechanism, but is consistent with the rapid step-growth reaction model. The RSGP model could be used, for example, to explain the effect of $P/(QM)$ on the structure, density, refractive index, surface energy, internal stress, and yield of deposition of hexamethyldisiloxane (HMDSIO) plasma polymer [1, 18]. During plasma polymerization of HMDSIO at a relatively low-input energy level, a very soft, polymerlike film was obtained, containing in its structure a significant amount of Si-CH_3 groups similar to the linear polymethylsiloxane. However, the plasma polymer obtained at a high-input energy level — $P/(QM) > 200$ MJ/kg — was extremely dense. At even higher values of $P/(QM)$ the obtained plasma polymer had a structure similar to an inorganic material composed of tetrahedrally bonded covalent elements of Si, C, and O [18]. In the intermediate ranges of $P/(QM)$

the obtained plasma polymer had inorganic characteristics which increased with increasing $P/(QM)$. This was demonstrated by the density change of HMDSIO plasma polymer. The density of HMDSIO plasma polymer was found to be between the density of organosilicon polymer, polydimethylsiloxane (0.98 $g.cm^{-3}$) and that of inorganic thin films, for example, silicon carbide or silicon dioxide (higher than 2.9 $g.cm^{-3}$), and the value of the density was uniquely controlled by $P/(QM)$. Gerstenberg et al. have also shown that films deposited by plasma polymerization of hexamethyldisilazane can be considered as intermediate materials between common polymers and amorphous solids [19]. During polymerization of tetrafluoroethylene it was found that fluorine-poor polymer film was deposited for high P/Q values, while fluorine-rich polymer was obtained for low P/Q [20].

The parameter $P/(QM)$ does not contain any indications on the effects of the geometries of the plasma reactors used in different studies, and it is therefore not correct to compare specific values of $P/(QM)$ between different reactors. The absolute value of $P/(QM)$ required for the deposition of a specific polymer film may vary with the type of employed reactor. A different energy input parameter has been proposed by Yeh et al. to include a geometric factor that takes into consideration the differences between various reactors [21]. The factor takes into account the volume of the plasma, the total reactor volume, and the changes in pressure due to the discharge. However, as reactor and plasma volumes are seldom reported in the literature, it is not possible to use that factor to compare plasma polymerization processes in different reactors.

Plasma polymerization occurs predominantly in the glow discharge. The reaction volume which is approximately the glow volume, Vol_g, is not always the same as the reactor volume, Vol_r. For example, in a bell jar reactor with parallel electrodes, the reaction volume is much smaller than the reactor volume. Therefore, the bypass ratio of monomer flow, which may be roughly proportional to the ratio Vol_r / Vol_g, should be taken into consideration when comparing plasma polymerization in different reactors [5]. Under typical conditions of plasma polymerization in a bell jar reactor, in which the plasma volume is only a small fraction of the total volume of the reactor, the conversion ratio of a monomer to a polymer is often less than a few percent, which means that the majority of the monomer is practically discarded.

Another parameter that reflects the material efficiency of the conversion of monomer to plasma polymer is the *specific conversion* parameter $DR/(QM)$, where DR is the nominal deposition rate on the substrate [1]. The plot of $DR/(QM)$ versus P/QM generally has three regions:

1. The *power-deficient region* at low values of $P/(QM)$, in which there is enough supply of monomer to use all the power; the power is the rate limiting factor and the specific conversion increases linearly with $P/(QM)$.
2. The *monomer deficient region* at high values of $P/(QM)$, in which the power supply is sufficient for conversion of the monomer into polymer

film and the monomer feed rate is the deposition-rate limiting factor; the specific conversion becomes constant or even decreases with increasing $P/(QM)$ in this region.

3. The *transition region*, which is intermediate between the first two regions where $DR/(QM)$ increases nonlinearly with $P/(QM)$.

Under certain plasma conditions the plasma polymerization can terminate in the gas phase, producing polymer powder. The powder formation generally occurs at intermediate pressures, high flow rates, but long residence times, high power, and conditions that reduce ion bombardment of the surface such as high frequency of the electric field sustaining the plasma [20].

The excessive formation of polymer powders occurs only under limited conditions, although powder formation has been observed in reactors of different designs and types of discharges and with various monomers. Powder formation was observed when organic vapor was introduced in a plasma of an inert carrier gas. A minimum inert gas pressure is required to obtain exclusively powders, and the yield of powder formation was found to be dependent on the inert gas used [5]. The aromatic hydrocarbons have been found to be most efficient in powder formation. Initiation of powder formation may result in the inclusion of particles in the polymer film, and this can be avoided by selecting the proper operational parameters and reactor design.

7.1.4 Examples of Plasma Polymerization

Plasma polymerization can also occur during reactive ion etching. The halocarbon gases used in reactive ion etching (RIE) polymerize easily if the excess halogen is consumed in the process. As etching is a large consumer of excess halogen, polymerization is commonly observed in RIE. This tendency to polymerization, to be discussed in the next chapter, is responsible for etch directionality and etch selectivity of silicon to silicon dioxide. The occurrence of either etching or polymerization in halogen-deficient RIE depends on the ion bombardment of the treated surface. Usually, etching is observed on ion bombarded surfaces, while polymer deposition occurs on grounded or nonbombarded surfaces [22].

Plasma polymerization has been used to prepare metal-containing polymers. Such materials have been prepared in fluorocarbon plasmas using a negatively biased metal electrode [23–25]. The metal of the electrode can be sputtered by bombardment with positive ions, but also removed by etching if it forms volatile fluorides. Silicon, germanium, molybdenum, and tungsten can be removed from the electrodes by both sputtering and chemical etching, while copper can be removed only by sputtering, because it does not form volatile fluorides [23]. The metals removed from the biased electrode can be incorporated in the matrix of the fluoropolymer formed by plasma polymerization. After exposure to atmosphere, the obtained structure has been characterized as a fine dispersion of metal oxides in a polymer matrix [23]. The metal oxides have been formed by the interaction with oxygen or water in the ambient. The composition of the metal-containing films can be controlled by changing the relative rates between etching and/or sputtering and deposition. Increasing the fluorine to carbon ratio in the precursor (e.g., using C_2F_6 instead of C_3F_8) results in an increase of the etching rate of the metal electrode and decrease of the deposition rate, causing a rise in the metal concentration in the polymer.

Composite materials of polymers containing metals have also been prepared by simultaneous or alternating plasma polymerization and metal evaporation [23, 26–28], or by direct plasma polymerization of metallorganic precursors, for example, titanium tetraisopropoxide [29]. In addition to the metals already mentioned, such composite materials also have been prepared with silver and gold.

Metallized insulators such as polymeric and ceramic materials are widely used in the appliance, automotive, and electronics industries. Metallization of nonconducting substrates is technically difficult because of structural incompatibility between the substrate and metallizing material, in terms of both chemical bonding and properties. Plasma preparation of metal-containing polymers has been used to deposit a transitional buffering film for use as a preplate in metallizing nonconducting materials, such as fiber reinforced plastic, PTFE (Teflon®), poly (ethylene terephthalate) (PET), nylon 66®, phenolic resin, and thermoplastic polyimide [30]. The process consists of depositing a graded film by sputter deposition of a suitable electrode material with simultaneous plasma polymerization of methane. The composition of the buffering films is changed from 100% organic to nearly pure metallic through the thickness of the films by modifying the deposition parameters during film growth. The deposition of the desired metallizing material is then performed by electroless deposition or electroplating.

Plasma polymerization is used to prepare films for corrosion protection of metals or water barriers for optical elements sensitive to humidity due to the possibility to prepare highly cross-linked, pinhole free films. This process is also used for preparation of gas separation membranes, reverse osmosis membranes for desalinization, abrasive-resistant coatings on softer materials, deposition of dry developable resist for lithography, and dielectric materials [1, 5, 20].

The deposition of plasma polymer film onto porous substrates has received special attention for preparation of gas separation membranes. Microwave plasma has been used to prepare separation membranes from monomer mixtures of hexamethyldisiloxane with methacrylate (HMDSIO/MMA) and styrene with vinyl acetate (S/VA) on porous silicone rubber [1].

Plasma polymerization was reported for vinyltrimethylsilane (VTMS), methyltrimethoxysilane (MTMOS), and vinyltrimethoxysilane (VTMOS) [17]. The obtained organosilicone polymers were found to be suitable as protective coatings for front surface mirrors made of Al or Ag used in various solar applications. The coatings also offered antistatic properties. Oxygen-containing plasma polymers of MTMOS and VTMOS have been found to have improved antistatic properties over the nonoxygen-containing polymer VTMS. Plasma polymerized VTMOS is used as an antireflective and abrasion-resistant coating on optical components made of polycarbonates. Plasma polymerized MTMOS coatings also provide abrasion resistance. Overcoats prepared by plasma polymerization are also employed to protect the aluminum surface in automotive reflectors [31].

Ultrathin solid, ionically conductive, polymer electrolyte membranes containing sulfonic ester groups were prepared by polymerization of methyl benzenesulfonate and octamethylcyclotetrasiloxane in a glow-discharge plasma, followed by a treatment with lithium iodide. Hybridization of this plasma polymer containing the lithium sulfonate groups with poly(ethylene oxide) resulted in the formation of a single lithium ion conductive film. The hybrid polymer electrolyte films were about 1 μm thick, pinhole free, and adherent to various substrates and showed ionic conductivities at 60 °C [32].

Plasma polymerization is also used to modify surfaces of biomaterials to reduce the interaction between biomaterial and the body fluids in contact with it. The chemistry of the biomaterial surface can be changed by plasma polymerization in plasmas of tetrafluoroethylene; ethylene; hexamethyldisiloxane; polar organic vapors, such as acetone, ethylene oxide, and methanol; and oxygen, affecting the manner in which proteins will adsorb at the surface. A list of plasma treatment of biomaterial polymers is shown in Table 6-3 of Chapter 6. It has been found that TFE-treated substrate becomes very retentive to albumin [33]. Coating polypropylene with plasma polymerized hexamethylcyclotrisiloxane (HMCTSO) resulted in reduced morphological changes of blood cells in contact with the coated surface [31].

Another example of the use of plasma polymerization for bioapplications is in packaging integrated circuits to be inserted inside the human body [34]. In this case, plasma polymerization of methane was used to deposit a thin adhesion-promoting layer on the circuit. The plasma polymerized layer was then covered with a thicker layer of Parylene-C®. The combination of the two layers has been found to provide excellent protection of the integrated circuit during long exposure to simulated body fluids [34].

An extensive list of applications for polymers produced by plasma polymerization can be found in references [5 and 6].

7.2 DEPOSITION OF INORGANIC FILMS

Deposition of coatings by PECVD is the most complex of all plasma surface treatment techniques. In-depth understanding is still limited, and the development of new deposition processes is mostly empirical. In thermal CVD the surface reactions are determined by thermodynamic considerations and controlled by the temperature of the substrate. Because most radicals react with surfaces, the composition of the films deposited by PECVD is determined to a large extent by the relative fluxes of all the species forming the film. This may be considered an advantage because it allows the formation of new materials, but it is also a disadvantage, as it makes control and reproducibility of composition difficult.

A large number of process parameters have to be controlled in plasma deposition, such as power, total pressure, flow rates of different gases, substrate temperatures, discharge frequency, electrode spacing, electrode materials, and reactor geometry. The effects of these parameters are interdependent and interact in determining material properties and deposition rates.

The importance of bombardment of the growing film with energetic particles is not yet clearly understood, and many deposition processes do not rely on this bombardment. It has nevertheless been shown that the temperature for silicon epitaxy can be reduced by ion bombardment during PECVD [35, 36] and that directional deposition can be obtained under ion bombardment [37, 38]. Special materials, such as diamondlike coatings, which will be discussed later, can be obtained only with the assistance of ion bombardment.

It should be emphasized that PECVD, similar to thermal CVD, by not being a line-of-sight process, enables deposition or surface treatment of complex shapes and sometimes closely packed specimens. Since this deposition method is suitable for a large variety of materials, only selected applications will be described in detail in this chapter. In depth discussions of PECVD applications for deposition of inorganic coatings used in microelectronic fabrication can be found in reference [39].

7.2.1 Amorphous Hydrogenated Silicon

Although amorphous hydrogenated silicon (a-Si:H) was already deposited by PECVD in 1966 by Chittick et al [40], the importance of a-Si:H became clear only after Spear and Le Comber showed in 1975 that amorphous silicon deposited from a plasma of silane can be made a *n*- or *p*-type semiconductor [41].

Amorphous tetrahedrally bonded silicon, obtained by sputtering or evaporation, contains many dangling bonds, as shown in Fig. 7-3, and thus many localized levels, rendering the material insensitive to doping. The amorphous silicon fabricated by Spear in 1975 using plasma decomposition of silane contained 10 to 30% hydrogen. The incorporated hydrogen terminates the dangling bonds (see Fig. 7-3) and reduces the localized state densities to the order of 10^{16} to 10^{17} $cm^{-3}.eV^{-1}$, thus providing sensitivity of material properties with respect to certain dopants. In other words, amorphous semiconductors became capable of the same *p-n* control required by electronic devices as crystalline semiconductors. The following year, Carlson and Wronski developed an a-Si:H solar cell [42]. In 1980 the first practical application of a-Si:H solar cell in an electronic calculator was achieved by Kuwano and Ohnishi [43].

Amorphous hydrogenated silicon is deposited from silane, usually at substrate temperatures of up to 200 °C. The temperature controls the amount of hydrogen incorporated in the film, which decreases with increasing temperature. The deposition is performed at pressures of 10 to 10^3 Pa (0.1 to 10 torr). The films can be doped by mixing small amounts of diborane (B_2H_6) or phosphine (PH_3) into the silane plasma obtaining correspondingly *p*-type or *n*-type silicon.

The deposition of a-Si:H from silane takes places through several reaction

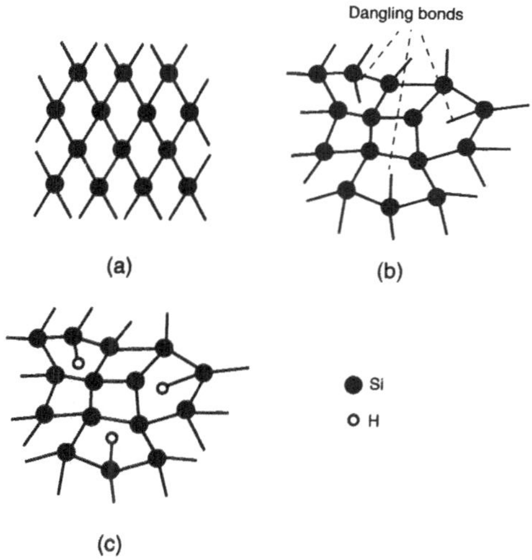

Fig. 7-3 Atomic bonds in silicon: (a) single-crystal Si; (b) amorphous Si; (c) hydrogenated amorphous Si.

steps. According to Veprek [44], in the first step, the silane is dissociated in the plasma into hydrogen and silene:

$$SiH_4 \rightarrow SiH_2 + H_2 \tag{7.2}$$

The dissociation is most probably induced by electron impact excitation of the vibrational degrees of freedom of the silane. The silene further reacts with silane to form disilane:

$$SiH_2 + SiH_4 \rightarrow Si_2H_6 \tag{7.3}$$

When sufficient disilane is formed in the plasma, the silene reacts with it to form also trisilane:

$$SiH_2 + Si_2H_6 \rightarrow Si_3H_8 \tag{7.4}$$

The rate controlling step of these three reactions is the rate of reaction for the formation of silene, described by Eq. (7.2). The disilane and trisilane decompose at the surface of the substrate to form the hydrogenated amorphous silicon:

$$Si_nH_{2(n+1)} \rightarrow n.\,\text{a-Si:H} + (n+1)H_2 \qquad n = 2,\ 3 \tag{7.5}$$

The rate limiting step of the di- and trisilane decomposition (Eq. (7.5)) is the rate of dehydrogenation of the surface of the growing film [44].

According to the described mechanism, the power density in the plasma controls the formation of di- and trisilane through Eq. (7.2). Ion impact on the surface of the film enhances its dehydrogenation and thus controls the deposition rate of the silicon. Ion impact also causes densification of the amorphous hydrogenated silicon.

Another suggested model explains the deposition of a-Si:H from silane, through the SiH_3 radical mechanism. The SiH_3 mechanism can be summarized by the following reaction steps:

$$SiH_4 \rightarrow SiH_3 + H \tag{7.6}$$

$$SiH_4 + H \rightarrow SiH_3 + H_2 \tag{7.7}$$

$$SiH_3 \rightarrow SiH_3 \text{ (adsorbed)} \tag{7.8}$$

$$SiH_3 \text{ (adsorbed)} \begin{cases} \rightarrow Si_xH_y\ (g) \\ \rightarrow \text{a-Si:H} \end{cases} \tag{7.9}$$

It is claimed in this model that at low-power discharges in pure silane, 98% of the silicon deposit is produced by SiH_3 radicals at low deposition rates [45]. According to the same model, at higher power levels, the deposition of silicon through SiH_3 radicals is replaced by deposition through SiH_2 radicals.

The reactions taking place in the silane plasma leading to the deposition of amorphous hydrogenated silicon are more complex and numerous than first described. A comprehensive description of the different reactions and their rate constants has been given in a model developed by Kushner for the deposition of a-Si:H by PECVD of silane [46].

While hydrogenated silicon was originally grown as an amorphous film and most of the applications are for a-Si:H, the material can also be prepared in microcrystalline form. For example, microcrystalline hydrogenated silicon films were deposited by RF PECVD of silane and hydrogen mixtures at substrate temperatures between 200 and 500 °C [47]. Two dilutions of silane in hydrogen, 1% and 10%, were studied. It was found that the growth rate, of the order of 100 Å/min., depends only slightly on substrate temperature and changed by about 17% in the studied temperature range. It was also found that the transition temperature between growth of amorphous and microcrystalline films decreased with increasing hydrogen dilution. For hydrogen dilutions of 1% and 10%, films deposited below 200 °C and 350 °C, respectively, were amorphous. Films deposited at higher temperature were found to be microcrystalline. The average grain size increased with substrate temperature due to decreasing hydrogen content in the films. The films deposited at a dilution of 10%, at temperatures below 350 °C, were found to have high photoconductivity and very low dark conductivity of $10^{-9}\Omega^{-1}.cm^{-1}$. The films deposited at the dilution of 1% had no photoconductivity and high dark conductivity of $10^{-3}\Omega^{-1}.cm^{-1}$, independent of the deposition temperature.

The manufacturing process of a-Si:H devices is simple because *p-i* and *i-n* junctions can be formed at the same time as the a-Si:H film. Thin film transistors have been built from phosphorus doped microcrystalline silicon. The phosphorus doped films were grown from plasmas of SiH_4 + 1%PH_3, diluted in hydrogen [48]. PECVD amorphous silicon has attained considerable importance as a new electronic material, for fabrication of solar cells or electronic devices such as thin film transistors or sensors [48, 49]. Hydrogenated amorphous silicon has become a common material for fabrication of solar cells. As the light absorption coefficient of a-Si:H is large, the solar cell thickness can be 1 μm or less [50]. Single-crystal silicon, by contrast, requires a thickness of approximately 300 μm for sufficient absorption of light. The photovoltaic conversion efficiency of solar light into electricity, which was initially extremely low at 2.4% for solar cells fabricated with amorphous silicon, attained more than 11% by 1984 [50].

The use of a-Si:H solar cell has grown to surpass the use of a conventional single crystal silicon solar cell in the field of consumer products. The industrial production of electronic and optoelectronic devices based on amorphous hydrogenated silicon approached 0.5 billion in 1989 [44]. Development of high-capacity, high-resolution flat computer or TV displays requires the integration of devices based on amorphous silicon films on large area substrates. Intrinsic and doped hydrogenated amorphous silicon are used in the fabrication of thin film transistors (TFT) for active matrix displays [51]. Attempts have also been made to apply the silicon films for sensing and printing devices [52]. Amorphous hydrogenated silicon films have been used to make image sensors on glass covered with a transparent conductor. The photosensor elements made of intrinsic a-Si:H were coupled to thin film transistors made by a-Si:H/a-SiC$_x$:H [53].

7.2.2 Silicon Nitride

Silicon nitride, widely used in microelectronics fabrication, for applications such as passivation of silicon devices of VLSI, gate dielectric, diffusion mask, or storage of electric charges [54], is another material deposited by PECVD. Silicon nitride prepared by PECVD is used as a gate dielectric in thin film transistors employed in the fabrication of flat panel displays [51]. Plasma deposited nitride and oxide are utilized as interlayer dielectric films between metallization levels. Silicon nitride is also used as an encapsulant of VLSI because the material is relatively hard and wear resistant and acts as an excellent sodium and moisture barrier. Because the silicon nitride is a very hard material, it protects the soft metal lines of microelectronic devices from scratches and other mechanical damage. Silicon nitride is used as an encapsulating coating for GaAs to prevent loss of arsenic during heat treatments. Another application is for etch masks in multilevel photoresist structures and for antireflection coatings for solar cells and photodetectors.

PECVD is extensively used for the preparation of silicon nitride because by this process the films can be deposited at temperatures below 400 °C, preventing thermal damage to the device on which it is deposited. The films deposited by PECVD are generally nonstoichiometric and often contain a significant amount of hydrogen. They are in fact hydrogenated amorphous silicon nitride and can be described by the formula a-Si$_x$N$_y$:H. The stoichiometry, referring to the Si/N ratio, hydrogen content, and the properties of the films, are dependent on the deposition conditions, that is, plasma parameters and substrate temperature.

Silicon nitride can be prepared from silane (SiH_4) and ammonia (NH_3), or silane and nitrogen using pure silane, or silane diluted in an inert gas. It can be grown in different types of reactors, under a variety of plasma conditions and substrate temperatures, as illustrated in Table 7-1. The properties of the films vary

TABLE 7-1 Various Conditions for Deposition of Silicon Nitride by PECVD (from [54], reprinted with permission)

System	f(Hz)	P_{rf}(W.cm^{-2})	T_s(°C)	P(T)	Gas Flow (sccm) SiH$_4$	Gas Flow (sccm) NH$_3$	G_R(Å/min)
A	13.56 M	0.04	120	0.3	?	?	170
			260				175
			380				150
B	13.56 M	0.38	100	0.5	22	80	290
			250				310
			370				320
C	13.56 M	0.02	100	0.32	17	120	159
			275				90
			320				—
			400				81
D	187.5 k	0.06	100	0.45	1400	—	140
			250				112
			350				141
E1	50 k	150 W	100	0.6	380	1130	100
			250	0.71	380	1705	89
		235 W	380	1.49	365	2735	223
			500	2.0	255	2945	218
E2	450 k	150 W	100	0.6	380	1130	203
			380	1.49	365	2735	212
			500	2.0	245	2945	204
F1	50 k	300 W	100	2.1	333	700	254
		250 W	250			1525	239
		300 W	380			2000	311
		200 W	600			2400	398
F2	450 k	300 W	100	2.1	333	700	288
			380			2000	309
		200 W	600			2400	369

accordingly, as shown in Table 7-2. For a large variety of deposition conditions, the hydrogen content was found to depend mainly on the substrate temperature and be almost independent of the Si/N ratio in the plasma mixture [54].

Because the reaction of silane with a nitrogen discharge is difficult to control, the reaction of silane with ammonia is more common. An earlier model [55] explained the formation of the amorphous hydrogenated silicon nitride by several reaction steps. In a plasma of silane and ammonia, electron impact causes the dissociation of silane and ammonia and produces primary reaction products as follows:

$$SiH_4 \rightarrow SiH_2 + H_2 \tag{7.10}$$

TABLE 7-2 Properties of Silicon Nitride Deposited in Conditions of Table 7-1 (from [54])

System	T_s (°C)	C_H (at. %)	Si/N Ratio		ρ (gr cm^{-3})	$\sigma \times 10^9$ (dyn cm^{-2})	n_r	E_g (eV)	E_R (Å/min)	t_r (Å)
			R_{BS}	R_M						
A	120	36.3	1.33	1.20	1.95	−1.63	1.98	4.12	3600	9860
	260	27.4	1.08	0.94	2.53	−2.17	2.02	4.63	170	8100
	380	18.9	0.90	0.87	2.91	−2.47	2.00	4.92	40	8700
B	100	33.1	0.78	0.97	2.17	1.73	1.83	5.13	8900	6640
	250	20.9	0.82	0.84	2.74	5.18	—	4.84	100	7180
	370	15.4	0.82	0.85	3.38	5.28	2.09	5.25	30	7400
C	100	38.8	—	—	2.49	0.59	1.78	5.07	3900	7150
	275	32.2	0.83	—	2.91	2.19	1.94	4.77	790	4960
	320	28.0	0.85	—	3.65	3.14	1.85	4.59	270	2830
	400	24.9	0.84	—	3.66	0.67	2.02	4.77	140	4840
D	100	28.9	0.78	—	—	−6.34	—	4.57	240	10760
	250	16.6	0.75	—	—	−12.28	—	4.93	20	8650
	350	20.1	1.04	—	—	−14.27	—	6.35	15	7900
E1	100	31.4	—	1.07	—	−9.85	2.03	3.87	560	5900
	250	24.6	—	0.87	—	−12.0	—	4.57	95	5400
	380	21.4	—	0.87	—	−9.41	—	4.50	65	5300
	500	15.7	—	0.86	—	−6.2	2.08	4.37	20	6230
E2	100	31.5	—	1.06	—	−4.84	2.05	—	870	9150
	380	19.8	—	0.91	—	−4.98	2.05	4.77	55	9130
	500	14.3	—	0.86	—	−4.34	2.12	4.75	20	8160
F1	100	32.0	1.08	0.81	—	−2.51	2.15	3.69	1890	4570
	250	25.7	—	1.16	—	−4.72	—	—	380	3820
	380	20.1	—	0.92	—	−8.03	—	4.45	50	9320
	600	12.3	0.81	0.80	—	−1.17	—	5.26	10	5170
F2	100	34.7	1.11	1.17	—	−3.07	—	4.68	1300	5200
	380	22.6	0.82	0.98	—	−7.18	—	4.78	50	4600
	600	11.1	—	1.16	4.32	0.54	—	5.36	6	9230

$$SiH_4 \rightarrow SiH_3 + H \tag{7.11}$$

$$SiH_4 \rightarrow Si + 2H_2 \tag{7.12}$$

$$SiH_4 \rightarrow SiH + H_2 + H \tag{7.13}$$

$$NH_3 \rightarrow NH + H_2 \tag{7.14}$$

$$NH_3 \rightarrow NH_2 + H \tag{7.15}$$

Radicals are also formed through exothermic disproportionation processes such as

$$H + SiH_4 \rightarrow SiH_3 + H_2 \tag{7.16}$$

$$H + NH_3 \rightarrow NH_2 + H_2 \tag{7.17}$$

Other species are formed through radical-molecule recombination reactions:

$$SiH_2 + SiH_4 \rightarrow Si_2H_6 \tag{7.18}$$

$$NH + NH_2 \rightarrow N_2H_4 \tag{7.19}$$

The SiH_x and NH_y radicals (mostly with $x = y = 1$) finally react to deposit the silicon nitride film. It has been postulated that SiH and NH are the primary free radicals promoting the deposition on the surface of the substrate [56]. The reactions described can be summarized as following:

$$SiH_4 + NH_3 + N_2 \rightarrow Si_xN_y{:}H_z \tag{7.20}$$

In a more recent model described by Smith [57], the growth of silicon nitride from ammonia and silane mixture takes place by the following steps:

- The ammonia and silane react in the plasma to form a "precursor" molecule, tetra-aminosilane, and hydrogen:

$$SiH_4 + 4NH_3 \rightarrow Si(NH_2)_4 + 4H_2 \tag{7.21}$$

- The precursor molecules deposit on the substrate as $Si(NH_2)_3$ after losing one amino group.
- Neighboring surface molecules condense to form silicon nitride and volatile ammonia.

At low-power density only silane reacts to form disilane. When the power is increased, the ammonia becomes activated and reacts with the silane to form the tetra-aminosilane, which deposits on the substrate and forms the silicon nitride film.

When nitrogen is used instead of ammonia, the film grows through formation of SiH_n radicals that stick to the surface of the substrate and react with N atoms. According to Smith [57], nitrogen requires about five times higher power density than ammonia to become completely activated and react fully with the silane. It was found that the films deposited from nitrogen had lower hydrogen content than those deposited from ammonia but poorer qualities.

Plasma deposited silicon nitride is amorphous, and because the films are formed from the hydrogen containing radicals, hydrogen is incorporated in them, as indicated by Eq. (7.20). In fact, hydrogen content has been measured to be over 35% for some plasma processes [58]. Equation (7.20) also indicates that silicon nitride is not stoichiometric ($x{:}y \neq 3{:}4$). The silicon to nitrogen ratio varies from 0.7 to greater than 4, but is usually close to 1. The silicon-to-nitrogen ratio is constant with film depth and depends mainly on the silane-to-ammonia or silane-to-nitrogen ratio in the precursor gases. The Si/N ratio also depends somewhat on the total gas flow rate, pressure, and power. It was found that nitrogen-rich silicon nitride films have better properties than do silicon-rich films [59].

Hydrogen plays an important role in determining film properties. The amount of incorporated hydrogen and its bonding determines the quality of the film. The hydrogen is evolved when the films are heated at temperatures above the deposition temperature, often leading to blistering and loss of adhesion. The etching rate

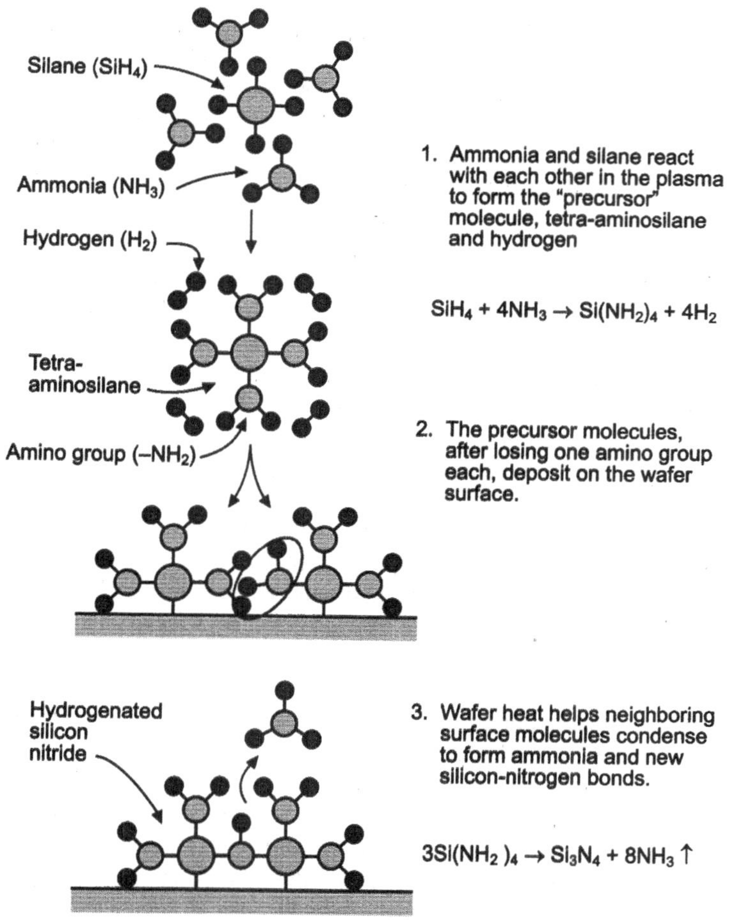

Fig. 7-4 Steps in the deposition of $a - Si_xN_y$:H from ammonia + silane plasma (from [57], reprinted with permission).

of PECVD deposited silicon nitride in aqueous hydrofluoric solutions correlates with the hydrogen concentration [60]. In addition, it has been found that excess hydrogen in plasma silicon nitride used for passivation can cause electrical instabilities in MOS devices.

It appears that large hydrogen content in PECVD silicon nitride films controls physical, chemical, and mechanical properties, while the optical and electrical properties were found to be dependent mainly on film stoichiometry (Si/N ratio). Best-quality films were obtained in nitrogen-rich films grown at temperatures above 350 °C [60].

The hydrogen concentration in the films can be reduced by increasing the hydrogen desorption rate during deposition through increase of the substrate temperature. At high pressure, temperatures in the range 380–400 °C are required to reduce the hydrogen in the film. These temperatures still reduce the hydrogen only to 16–18% levels, compared with typical values of 12–14% obtained at low pressure (0.7 torr) and a lower temperature of 300 °C. It seems, therefore, that the way to deposit high-quality silicon nitride films useful for practical applications is by choosing low pressure and low temperature [61].

It has also been observed that the amount of hydrogen incorporated in silicon nitride films deposited by PECVD from silane-nitrogen or silane-ammonia mixture could be reduced by adding argon to the plasma [62]. The reduced amount of hydrogen in the films was explained by the sputtering of hydrogen molecules adsorbed on the surface of the growing film through bombardment with the argon ions.

The density of plasma deposited silicon nitride ranges from 2.1 and 3.1 $g.cm^{-3}$ and depends on all the major deposition variables. The observed resistivity of plasma deposited silicon nitride ranges from 10^5 to greater than 10^{20} Ω.m. The resistivity is a function of deposition parameters including power, pressure and reactant gas ratio. Breakdown fields are usually between 1 and 6×10^7 $V.cm^{-1}$, depending on film composition. Films that contain excess silicon have low breakdown fields. The dielectric constant, measured at 1 MHz, is usually between 6 and 8. However, the dielectric constant can range between 4 and 13, depending strongly on the silicon and hydrogen content of the film (e.g., [63]).

PECVD silicon nitride films are characterized by stresses which range from 10^{10} $dyn.cm^{-2}$ compressive to 10^{10} $dyn.cm^{-2}$ tensile. The type (tensile or compressive) and values of the stresses are important as they can affect the structure coated with the silicon nitride. Films deposited at low frequency usually have compressive stresses, while films formed at a high frequency have tensile stresses. Annealing at temperatures as high as 700 °C has no effect on the film stress even though hydrogen is lost during the anneal. Stresses depend on the substrate temperature, gas composition, power, and frequency. It has been found that under certain deposition conditions, films deposited in a plasma excited at 500 kHz have higher stresses and higher charge density than do films grown at 13.56 MHz [57]. The difference in quality was explained by implantation of H^+ ions at the lower frequency [57] at which the substrate has a larger self-bias.

Silicon nitride and silicon oxynitride (SiO_xN_y) have also been prepared by the remote PECVD (RPECVD) technique. Nitrogen and oxygen sometimes mixed with an inert gas are dissociated in an RF plasma [64] or in a microwave plasma [65]. The chemically active nitrogen and oxygen species are transported into the reaction chamber, where they mix with silane that is admitted outside the plasma region. The silane reacts heterogeneously on the substrate with the reactive species to form the nitride (or oxynitride) film, with hydrogen content much smaller than in the films deposited by conventional PECVD [64]. $Si_xN_yH_z$ films grown with the remote microwave plasma, on substrates at temperatures in the range 70–350 °C, had a $x{:}y$ ratio near 3:4, but contained 2–30 at% hydrogen [65]. Films grown by a similar process, but using a 13.56 MHz RF plasma, contained almost no hydrogen [64].

ECR plasma has also been used to grow silicon nitride and silicon oxynitride by an RPECVD process. Nitrogen and/or oxygen were inserted through the ECR plasma source, while silane, diluted to 25% in helium, was introduced next to the substrate, outside the plasma. The deposition was performed on substrates at temperatures of 85–110 °C. The composition of the films could be controlled by varying the flow rate ratio for oxygen/nitrogen [66].

7.2.3 Silicon Dioxide

Silicon dioxide (SiO_2) deposited by PECVD is used as an interlayer dielectric film between metallization layers in VLSI. Silicon dioxide films have been deposited from a variety of gas mixtures using different plasmas and process parameters. Silicon dioxide has been deposited by PECVD mainly from plasmas of silane with nitrous oxide, or silane, nitrous oxide, and oxygen, at substrate temperatures between 200 °C and 400 °C. The optimal deposition temperature has been found to be a function of the precursor and deposition technique.

High-quality silicon dioxide films, comparable to those obtained by high-temperature CVD (700–800 °C), have been grown by PECVD from silane highly diluted in helium and nitrous oxide [67]. Good-quality oxides have also been obtained from a plasma of silicon tetrachloride ($SiCl_4$) and oxygen, at substrate temperatures of 200 °C [68].

Silicon dioxide films have also been deposited by ECR plasma of 10% silane in argon and oxygen, at temperatures of 150–320 °C. Good electrical properties were obtained for the specific system in films deposited at substrate temperatures higher than 285 °C [69]. The quality of the films has been found dependent on the flow rate; at increased flow rates nonhomogeneous porous films were obtained.

Another precursor for PECVD of SiO_2 is tetraethoxysilane (TEOS). Silicon dioxide has been deposited from TEOS and oxygen [70], nitrous oxide, or ozone [71] at a variety of frequencies, from 150 kHz to 2.45 MHz. In spite the fact that the precursor contains oxygen, films deposited from a plasma of only TEOS, in the absence of an oxidant, have been found to contain up to 18% carbon. When oxygen is added to the plasma as an oxidant, the carbon content can be reduced to about 1%, while when nitrous oxide (N_2O) is used as oxidant, the films incorporate about 2% nitrogen [72]. The quality of the films deposited from TEOS and oxygen has also been found to depend on the excitation frequency. Better-quality films were obtained at lower frequencies (150 kHz) than at high frequency (14 MHz) [73].

Remote PECVD is often used to deposit silicon dioxide films [64]. Active species are formed in the plasma and flow away from the discharge to the substrates. Since the discharge and the substrates are separated, radiation damage to the substrate is avoided.

Silicon dioxide has been grown directionally in a 14-MHz plasma from mixtures of oxygen and tetraethoxysilane. The liquid TEOS was carried into the reactor by helium flowing through the evaporator. The growth directionality was found to be a function of the O_2:TEOS ratio. The quality of the films could be improved by adding a small amount of NF_3 [38]. The addition of NF_3 also controlled the taper of the growth profile on the sidewall of a coated trench.

Doped oxides can be prepared by adding phosphine or diborane to the plasma mixture. Phosphine is used to produce a phosphorus-doped silicon dioxide (phosphosilicate glass, or P-glass). Diborane and phosphine are both added to form borophosphosilicate glass, which is useful because it flows at low temperatures and can be used for planarization of VLSI. The doped films usually contain between 2 and 10 wt% phosphorus or boron [38].

Similar to silicon nitride, plasma deposited silicon dioxide is amorphous and usually contains hydrogen. The total hydrogen concentration varies between 2 and 9 at%, with most samples containing about 5 at% hydrogen. Silicon dioxide films deposited from silane and nitrous oxide contain hydrogen bonded as SiOH, H_2O, and SiH. Some of the hydrogen is trapped or adsorbed in the oxide. The refractive index of plasma deposited silicon dioxide is usually between 1.45 and 1.55 and depends on the preparation conditions. For example, it has been found that films deposited with a nitrous oxide-to-silane ratio equal to 4.5 had a high refractive index and were silicon rich, with a silicon-to-oxygen ratio of 0.89. Films deposited with a nitrous oxide to silane ratio equal to 15 had a lower refractive index, 1.5, and a silicon to oxygen ratio of 0.53 [60].

The breakdown strength of silicon dioxide is usually between 2 and 6×10^6 V/cm, although the maximum strength can be as high as 8 to 10×10^6 V/cm. The plasma deposited silicon dioxide films have a low compressive stress, usually between 1 and 3×10^9 dyn.cm^{-2}. The stress of phosphorus doped films is very close to zero when the phosphorus concentration is between 5 and 8 wt%.

The dielectric constant of plasma deposited silicon oxide is also affected by the preparation conditions. For silicon dioxide deposited from silane and nitrous oxide in an RF plasma, the composition of the films could be varied by changing the SiH_4:N_2O ratio between 1:1 and 1:10 resulting in a change of the dielectric constant between 3.8 and 10 [63]. Amorphous silicon nitride deposited from ammonia and silane had dielectric constants between 8 and 15 (at 1 kHz).

The deposition rate of silicon dioxide is enhanced with increasing RF power and substrate temperature. For deposition from silane-nitrous oxide plasmas, the deposition rate increases when the silane concentration and total flow are raised. The deposition rate seems to be independent of total pressure and the nitrous oxide/silane ratio.

7.2.4 Silicon Carbide

Silicon carbide was among the first coatings deposited by PECVD [74]. Plasma deposited silicon carbide can be amorphous or microcrystalline. It has been deposited from mixtures of silane (SiH_4) with methane (CH_4), acetylene (C_2H_2) [75], or ethylene (C_2H_4) [76, 77] or by plasma decomposition of tetramethylsilane ($Si(CH_3)_4$) [78].

Plasma deposited silicon carbide films are also hydrogenated amorphous films, similar to PECVD prepared silicon, silicon dioxide, or silicon nitride. Silicon carbide, which is often described as a-SiC:H, can be best expressed by the formula a-Si_xC_y:H. The silicon-to-carbon ratio in the plasma deposited films varies from 0.1 to 4 (Si_4C to SiC_9). The composition of the films usually depends on the ratio of reactants and is nearly independent of substrate temperature [74], which can vary from 25 °C to 600 °C.

During PECVD of hydrogenated amorphous silicon carbide from mixtures of silane with hydrocarbon gases, it was found that at fixed molar ratios of plasma gas, the composition of the films depends strongly on the nature of the hydrocarbon [79]. The same film composition can be obtained from mixtures of silane with different hydrocarbons at different ratios. Optical bandgap, refractive index, and mechanical abrasion resistance were found to be dependent on the composition of the film and not on the precursors used.

The nonstoichiometric amorphous hydrogenated silicon carbide can be silicon or carbon rich. It has been found that the physical properties of PECVD silicon-rich a-SiC:H are only slightly affected by the power density of the plasma, whereas the properties of the carbon-rich films are strongly dependent on power density [80].

Plasma deposited silicon carbide films have resistivities of about 10^{12} Ω.cm, refractive indexes between 1.96 and 2.6, and compressive stresses of 0.2 to 5×10^9 dyn.cm^{-2} [78]. The properties of the films, such as the optical gap, can be adjusted by varying the composition of the plasma mixture. By adding diborane (B_2H_6) or phosphine (PH_3) to the plasma gas, the films can be doped to obtain *p*-type or *n*-type films correspondingly. The films can be used in optoelectronic applications, such as photoreceptors, photovoltaic cells, and thin film light-emitting diodes.

The deposition rate of silicon carbide depends on power and gas flow. At low power, the deposition rate was found to increase with power and to be independent of flow, while at high power, the rate becomes independent of power and is transport limited, depending strongly on gas flow. The deposition rate is determined be the generation of active species in the discharge and is independent of substrate temperature.

Polycrystalline (microcrystalline) silicon carbide films have also been deposited by RF PECVD from tetrafluorosilane (SiF_4), tetrafluoromethane (CF_4), and hydrogen mixtures. Transmission electron microscopy showed that the films had the α-SiC structure [81]. Hydrogenated microcrystalline silicon carbide has also been prepared by ECR PECVD [76].

Hydrogenated amorphous silicon carbide is gaining importance for fabrication of solar cells and for electronic devices. High-efficiency solar cells have been prepared by depositing microcrystalline silicone carbide films on polycrystalline silicon using an ECR plasma. A conversion efficiency of 15.4% has been obtained. Using such a cell as the bottom component in a tandem-type solar cell with an amorphous silicon, an efficiency of 16.8% was obtained [82]. It has been found that a DC bias applied to the substrate increased the deposition rate and improved the photoconductivity of the films. Amorphous hydrogenated silicon carbide deposited by PECVD is also used as the topmost passivation layer in a multilayer thin film device of a thermal ink jet printhead [83].

7.2.5 Diamondlike Carbon

Diamondlike carbon or DLC films are metastable amorphous materials, which may include a microcrystalline phase. The material is characterized by extreme hardness, in the range 1000–3000 $kg.mm^2$, and a generally low-friction coefficient, between 0.01 and 0.28. DLC films typically have high optical transparency over a wide spectral range, high electrical resistivity (up to 10^{16} Ω.cm), and chemical inertness to both acids and alkalis. A problematic attribute of DLC is the very high internal stresses of the order of 1 GPa. These disordered forms of carbon are called "diamondlike" due to their extreme hardness and significant optical gap.

The deposition and characterization of DLC films have been reported upon extensively in the literature and can be found in summary form in reviews [84, 85] and conference proceedings such as [86, 87]. Wide interest in the deposition and characterization of DLC films stems from the attractive properties of these films as enumerated earlier and from the fact that very smooth films can be deposited uniformly over wide areas on a large variety of substrates, making possible commercial use of DLC films in numerous applications.

Since first deposited by Aisenberg and Chabot in 1971 [88], DLC films have been prepared by a variety of methods, including DC or RF plasma assisted CVD, sputtering, and ion beam deposition, using different carbon bearing, solid or gaseous, source materials. Different forms of PECVD constitute the main method for deposition of diamondlike carbon films. Among the PECVD methods, parallel plate RF reactor systems are the most common type employed, but microwave and ECR plasmas are also used.

Any hydrocarbon with sufficient vapor pressure can in principle be used as the precursor for PECVD of DLC films. Among those precursors are acetylene, benzene, butane, cyclohexane, ethane, ethylene, hexane, izopropane, methane, pentane, propane, and propylene. Hydrogen has on occasion been added to the hydrocarbon precursor; however, it appears that this addition is not required for the deposition of DLC films and does not significantly alter their properties.

In addition to the large variety of precursors, the deposition conditions also cover large ranges of parameters:

- RF power: 15–1000 W; the more significant parameter is the power density, but it is not always possible to determine its value from the published results. It seems that power densities of order of 0.1–0.5 $W.cm^{-2}$ are the most suitable for obtaining films with diamondlike properties
- Negative substrate bias: up to 1500 V DC

- Pressure: 0.006–7.5 torr
- Substrate temperature: room temperature − 250 °C.

All methods for the deposition of DLC are nonequilibrium processes characterized by the interaction of energetic ions with the surface of the growing film. PECVD of DLC has to be done on negatively biased substrates to provide bombardment of the growing film with energetic ions. The metastable structure of DLC films most likely originates from the thermal and pressure spikes produced by impinging energetic species at the growth surface. The energy of a 100-eV ion is dissipated in a very small near-surface volume of the order of several angstroms, resulting in power dissipation of the order of $W.cm^{-3}$. Thermal spikes of 3283 °K and pressures of 1.3×10^{10} Pa (1.2×10^{5} atm) for about 7×10^{-11} sec. have been calculated for impinging 100-eV ions [89]. The duration of the spikes is much longer than the vibrational period of 2.6×10^{-14} sec. for diamond and is thus well beyond that required to allow bonding. The metastable DLC structure formed is conserved due to extremely high quenching rates of the thermal spikes.

Radicals appear to be the primary factor in the growth of DLC films. Addition of a radical scavenger to the hydrocarbon gas can completely suppress growth of DLC films [90]. Ion bombardment is another process important to the growth and properties of such films. The radicals and ions produced by the dissociation of the precursor in the plasma are further fragmented upon surface impact, after which the near-surface excited region is quenched by the underlying colder substrate. Higher growth rates and harder films are obtained on the powered electrode of an RF PECVD apparatus than on the grounded electrode. Vandentop et al. [90] explained this effect as ion bombardment leading to more complete fragmentation of the hydrocarbon species into ever-more hydrogen-deficient radical species, and thus increasing the sticking coefficient on the bombarded surface.

The deposition rate of DLC increases with decreasing temperature and increasing ion bombardment. It was shown that the growth of the DLC film takes place through the interaction between the substrate surface and a physisorbed hydrocarbon layer [91]. The growth rate is therefore determined by the impingement rate of plasma species and their residence time on the physisorbed surface layer. The latter decreases with increasing temperature. This mechanism explains the increase of the growth rate with decreasing temperature and increasing ion flux [91]. The rate of deposition was found to increase more than one order of magnitude for identical deposition condition when methane is replaced by benzene. The change in deposition rate was much higher than the ratio of the masses of the precursor molecules, and it was explained by higher ion fluxes in the benzene plasma, caused by lower ionization energy of benzene [92].

PECVD prepared DLC films contain a significant amount of hydrogen. When using different deposition methods, the hydrogen content in DLC films was determined to be from less than 10 to 60%. DLC films are essentially an amorphous network of sp^3 (diamondlike), sp^2 (graphitelike), and even sp^1 coordinated carbon atoms in a disordered network. The ratio between the carbon atoms in the different coordinations was found to be a strong function of the total hydrogen content of the films. Hydrogen is also important for obtaining a wide optical gap and high electrical resistivity, as it passivates the dangling bonds in the amorphous structure [89]. This passivation removes midgap defect states and stabilizes the random network, preventing its collapse into a graphitic phase.

The properties of DLC films deposited by PECVD have been found to be generally insensitive to the chemical precursor employed, but to be strongly dependent on the energy of impacting ions. Koidl et al. [92] found that the properties of DLC films deposited from nine different hydrocarbon species were independent of the precursor if the negative substrate bias was in excess of 100 V DC.

DLC may be described as a disordered structure of clusters of aromatic rings interconnected by sp^3 carbon bonds [93]. While diamond is a well-defined material with fixed properties, the properties of DLC cover a wide range of values between those of diamond, graphite, and hydrocarbon polymers.

The properties of DLC films can be a function of several plasma parameters, the dominant influence being attributed to the substrate bias; it should be noted, however, that in most instances this characteristic is only indirectly controlled in deposition of DLC. Variation of substrate bias in RF PECVD deposition systems is frequently achieved by changing the RF power (P_{RF}) and/or the pressure, p, in the reactor. According to Catherine [91], the self-bias of the substrate, V_B, is related to these two parameters by

$$V_B \propto \left(\frac{P_{RF}}{p} \right)^{1/2} \tag{7.22}$$

At the relatively high pressures used in PECVD of DLC, the ions are scattered by collisions and reach the substrate with a distribution of energies lower than the applied bias. The average impact energy of ions bombarding the growing film in the RF plasma depends on the DC bias and on the pressure. Bubenzer et al. [94] found the relation

$$E \propto \frac{V_B}{p^{1/2}} \tag{7.23}$$

Thus the properties of the DLC films can be varied by adjusting the bias directly or by adjusting the growth pressure. It is claimed [92] that a substrate bias higher than 100 V DC is required to obtain hard films. However, this lower bias limit value can depend on the pressure during deposition because high-pressure operation will reduce the average energy of species impinging the growth surface. If the impact energy is too low, "polymerlike" films can be obtained [92]. Conversely, at sufficiently low pressures, this lower bias limit is relaxed, and hard wear-resistant films can be obtained at biases below 100 V DC [95].

The properties of DLC films can also be affected by substrate temperature during deposition. It was reported [84, 85] that DLC's optical gap decreases with increasing substrate temperature, with a rapid decrease above 250 °C. Above 325

°C the electrical resistance of the films decreases strongly (10^{-2} Ω.cm), and the films soften significantly. Deposition of DLC is therefore generally performed at substrate temperatures below 300 °C. Low-deposition temperature may lead to the formation of a polymerlike film, or polymerlike fraction embedded in a hard DLC layer. The deposition temperature has a significant impact on the hydrogen content of DLC films and on their thermal stability and better thermal stability has been observed for DLC film deposited at higher temperatures [96].

The properties of DLC can be tailored by adjusting the deposition parameters, such as substrate bias, temperature, and power density. High hardness and chemical resistance of DLC films make them good candidates for use as wear-resistant coatings on metals on optical and electronic components. The use of DLC is especially attractive in applications where it is required that the thickness of the protective film should be less than 50 nm thick, as for example for magnetic recording media; however, thicker DLC films are also used in various applications [97].

7.2.6 Diamond

Diamond films are to be distinguished from diamondlike carbon discussed in the previous section. The diamond films are polycrystalline materials, with crystallites up to tens of microns in size, having diamond's crystal structure, while DLC films are essentially amorphous materials without any long-range order. Diamond is characterized by fascinating and outstanding properties. It has the highest atom number density, and as a result of this density and strong covalent bonding, it has the highest hardness and elastic modulus of any material. It has a thermal conductivity 10 times higher than that of pure copper. Diamond has semiconductive properties being a very wide bandgap semiconductor (5.5 eV) and having a high breakdown voltage (10^7 V.cm^{-1}) [98]. It is also characterized by chemical inertness.

Due to its attractive properties, efforts have been invested to grow synthetic diamonds. While industrial macroscopic diamond crystals have been grown since 1955 at high temperatures (up to 3500 °C) and very high pressures (a few hundreds kbar) [99], progress has been made in recent years in growing diamond films by chemical vapor deposition. Diamond CVD is performed at atmospheric or lower pressures.

The CVD methods for growing diamond films are based on the production of high concentrations of atomic hydrogen in the gas phase. It is believed that during the CVD process, both diamond and nondiamond carbon phases (e.g., graphite) deposit on the substrate and the atomic hydrogen etches away the nondiamond phases, promoting the growth of the diamond. Diamond is also etched by atomic hydrogen, but at a rate 50 times lower than that of nondiamond carbon [99]. It is also believed that atomic hydrogen initiates gas phase chemistry and stabilizes the tetrahedral sp^3 carbon on the surface of the growing film.

PECVD techniques are among the CVD methods for deposition of diamond films. Of the different PECVD processes, microwave and ECR plasmas are those adopted for deposition of diamond films. Microwave plasma has become one of the standard methods for deposition of polycrystalline diamond films [100, 101]. The microwave plasmas used for the diamond deposition are generally sustained at relatively high pressures of 10 to 200 torr. At these pressures the microwave discharge is generally concentrated over a small area, limiting the area on which uniform films can be deposited. This limitation can be overcome by using ECR plasmas at lower pressures of about 0.1 torr [102]. The ECR plasma provides a uniform flux of radicals and high-density flux of ions over large areas, enabling uniform diamond deposition over such areas. Specially designed nonmagnetic microwave plasmas have recently been developed allowing the simultaneous deposition of diamond over two 8-inch wafers [103].

The deposition is performed at substrate temperatures of 850 to 950 °C. When the substrate is in direct contact with the plasma, it is heated by the plasma and its temperature is determined by the microwave power. In some microwave plasma tools the microwave power is therefore limited by the heating of the substrate, which is thus restricting the growth rate of the film. To overcome this power limitation problem and increase the growth rate, deposition of diamond films has been performed with pulsed microwave plasma [104]. In other configurations, the substrates may have to be heated independently of the plasma.

The deposition is performed usually from methane or other hydrocarbons diluted in hydrogen, usually to less than 1%. At higher hydrocarbon concentrations, the films contain increased fractions of graphite or amorphous carbon. To obtain high-quality diamond films, the substrates have to be heated at high temperatures, usually above 800 °C. Although the plasma heats the substrates and dissociates the hydrocarbons, in the deposition of diamond its main role is to produce atomic hydrogen. The plasmas are therefore sustained at conditions that create high concentrations of atomic hydrogen. Theoretical models of diamond growth claim that the deposition takes place through methyl radicals [105] or through formation of acetylene species that deposit and rearrange on the substrate [106]. It is not yet clear which one of the two mechanisms is responsible for the actual deposition.

Addition of oxygen to the plasma shows beneficial effects on the quality of the films [107]. It is added to the gas mixture either as molecular oxygen or as a compound with carbon and/or hydrogen. Atomic oxygen generated in the plasma enhances the etching of non-diamond carbon phases [108], reduces the gas phase hydrocarbon [109] and improves the growth rate [110] and quality [108, 110] of the diamond films. Oxygen also reduces the substrate temperature at which good diamond films are obtained. Diamond films have been grown from a gas mixture to which 0.5 to 1% water vapor were added [111] or using a remote microwave plasma of 5% CH_4 + 5% O_2 in H_2, at a pressure of 10 torr and substrate temperature of 600 °C. In the latter case the substrates were RF inductively heated [112]. Diamond deposition was also performed from hydrogen diluted in carbon monoxide on a substrate independently heated to 650–750 °C. In this case, the substrate had to be biased positively to obtain good-quality diamond [113].

Diamond films can be grown epitaxially on diamond substrates, while deposition on other substrates requires special substrate preparation to promote diamond nucleation. The substrates are generally polished or abraded with diamond powder, or even sprayed with fine diamond particles before insertion in the deposition reactor. It is claimed that diamond film deposition starts by reduction of surface oxides and the formation of a carbide or carbidic layer.

High nucleation density is obtained by the surface preparation of the substrate, and as a result the diamond films are polycrystalline. The films grow as faceted crystallites having well-defined crystallographic habits. They appear as octahedral, cubo-octahedral, or twined particles truncated by the substrates, oriented normal to the growth surface. The films produced are therefore very rough, and this is one of the reasons limiting their applications. The growth shape and crystallographic orientation of the diamond crystallites is a function of growth condition and gas composition. Oxygen addition can change the crystallographic orientation and crystallite size [114].

Deposition of wear-resistant diamond could improve or create new cutting tools with improved performance or provide wear-resistant coatings on optical lenses and bearings. Highly transparent diamond films would be an ideal window material for electromagnetic radiation, ranging from X rays to microwaves. Due to its high thermal conductivity, diamond can easily outperform conventional copper or silver in heat sinks [99]. Free-standing stiff diamond membranes can be used for X-ray lithography masks needed for fabrication of higher-density VLSI. The combination of high thermal conductivity with extreme hardness and chemical inertness render CVD diamond ideal for fluid flow heat exchangers and as substrates for electronic circuits. Diamond films can be doped, and having a large bandgap of 5.5 eV, they are an interesting material for high-temperature passive or active electronic devices, for example, thermistors or transistors operating at temperatures of more than 500 °C [99]. Potential electronic applications of diamond also include high-temperature devices, millimeter-wave traveling wave amplifiers, and picosecond high-voltage electro-optic devices [98].

Practical applications of diamond films are dependent on achieving good adhesion, smoothness, crystallographic perfection, or single-crystal epitaxy. While optically smooth diamond coatings have been obtained, other properties have yet to be attained to further new uses of the diamond films.

7.3 QUESTIONS

1. Explain the differences between plasma polymerization and plasma induced polymerization.
2. a. What is the main plasma parameter that affects polymerization in the plasma?

   b. What is the advantage of polymerization in a plasma as compared to conventional polymerization?

3. Search the recent literature for at least three different studies of plasma polymerization (as a general term) and summarize the studies and the possible applications.
4. Describe the common features of PECVD deposited inorganic films.
5. Search the recent literature for studies of plasma deposited amorphous silicon. Choose three studies on the effects of plasma parameters on the film properties and critically discuss the findings.
6. Repeat question 5, but for silicon nitride and silicon carbide.
7. Search the recent literature for studies of a material not discussed in the book. Critically discuss the findings.

7.4 REFERENCES

[1] Kramer, P. W., Y. Yeh, and H. Yasuda, *J. Membr. Sci.*, 46: 1 (1989).

[2] Yasuda, H., In *Thin Film Processes*, eds. W. Kern and J. L. Vossen, p. 361. New York: Academic Press, 1978.

[3] Shen, M., and A. T. Bell, In *Plasma Polymerization*, eds. M. Shen and A. T. Bell (ACS Symp. Series, 1979), Vol. 108, p. 1.

[4] Bell, A. T., In *Topics in Current Chemistry, Plasma Chemistry III*, eds. S. Veprek and M. Venugopalan, p. 43. Berlin: Springer-Verlag, 1980.

[5] Yasuda, H. K., *Plasma Polymerization*. New York: Academic Press, 1985.

[6] d'Agostino, R., ed., *Deposition, Treatment and Etching of Polymers*. New York: Academic Press, 1990.

[7] Poll, H. U., M. Artz, and K. H. Wickleder, *Eur. Polym. J.*, 12: 505 (1976).

[8] Haller, I., and P. White, *J. Phys. Chem.*, 67: 1784 (1963).

[9] Turos, A., and O. Meyer, *Nucl. Instrum. Methods Phys. Res.*, B4: 92 (1984).

[10] Smolinsky, G., and M. J. Vasili, *J. Macromol. Sci. Chem.*, 10: 473 (1976).

[11] Kobayashi, H., M. Shen, and A. T. Bell, *J. Macromol. Sci. Chem.*, 8: 1345 (1974).

[12] Morita, S., T. Migutani, and M. Ieda, *Jpn. J. Appl. Phys.*, 10: 1275 (1971).

[13] Bradley, A., and D. Fales, *Chem. Technol.*, 1: 232 (1971).

[14] Carmi, U., A. Inspector, and R. Avni, *Plasma Chem. Plasma Process.*, 1: 233 (1981).

[15] Morosoff, N., H. Yasuda, E. S. Brandt, and C. N. Reilley, *J. Appl. Polym. Sci.*, 23: 3449 (1979).

[16] Lopez, G. P., and B. D. Ratner, In *Plasma Polymerization and Plasma Interactions with Polymeric Materials*, ed. H. K. Yasuda, p. 511. New York: John Wiley & Sons, 1990.

[17] Yasuda, H., and T. Hirotsu, *J. Polym. Sci., Polym. Chem. Ed.*, 16: 743. (1978).

[18] Park, S. Y., and N. Kim, In *Plasma Polymerization and Plasma Interactions with Polymeric Materials*, ed. H. K. Yasuda, p. 91. New York: John Wiley & Sons, 1990.

[19] Gerstenberg, K., *Colloid Polym. Sci.* (West Germany), 268: 345 (1990).

[20] Morosoff, N., In *Deposition, Treatment and Etching of Polymers*, ed. R. d'Agostino, p. 1. New York: Academic Press, 1990.

[21] Yeh, Y., I. Shyy, and H. Yasuda, *J. Appl. Polym. Sci., Appl. Polym. Symp.*, 42: 1 (1988).

[22] Coburn, J. W., and H. F. Winters, *J. Vac. Sci. Technol.*, 16: 371 (1979).

[23] Kay, E., J. Coburn, and A. Dilks, In *Topics in Current Chemistry, Plasma Chemistry III*, eds. Veprek, S. and Venugopalan, M., p. 1. Berlin: Springer-Verlag, 1980.

[24] Biederman, H., R. P. Howson, and I. McCabe, In *Proc. 6th Int. Conf. Ions and Plasma Assisted Technol. (IPAT 87)* (1987), p. 152.

[25] Koeberle, H., M. Grischke, F. Thieme, C. Benndorf, and R. Memming, *Surf. Coat. Technol.*, 39/40: 275 (1989).

[26] Biederman, H., I. Chudacek, L. L. Martinu, J. J. David, and S. Nespurek, *Vacuum*, 39: 13 (1989).

[27] Heilmann, A., C. Hamann, G. Kampfrath, and V. Hopfe, *Vacuum*, 41: 1472 (1990).

[28] Stulik, P., H. Biederman, D. Slavinska, A. Fejfar, I. Chudacek, and P. Mackus, *Int. J. Electron.*, 70: 509 (1991).

[29] Kagami, Y., T. Yamauchi, Y. Osada, and I. Yoshizawa, *J. Appl. Phys.*, 68: 610 (1990).

[30] Yasuda, H., T. J. O'Keefe, D. L. Cho, and B. K. Sun, In *Plasma Polymerization and Plasma Interactions with Polymeric Materials*, ed. H. K. Yasuda, p. 209. New York: John Wiley & Sons, 1990.

[31] Wrobel, A. M., and M. R. Wertheimer, In *Deposition, Treatment and Etching of Polymers*, ed. R. d'Agostino, p. 163. New York: Academic Press, 1990.

[32] Ogumi, Z., Y. Uchimoto, Z. Takehara, and F. Foulkes, *J. Electrochem. Soc.*, 137: 29 (1990).

[33] Hoffman, A. S., In *Plasma Polymerization and Plasma Interactions with Polymeric Materials*, ed. H. K. Yasuda, p. 341. New York: John Wiley & Sons, 1990.

[34] Charlson, E. J., R. Sabeti, E. M. Charlson, and H. K. Yasuda, *Mat. Res. Soc. Proc.*, 154: 91 (1991).

[35] Donahoe, T. J., and R. Reif, *J. Appl. Phys.*, 57: 2757 (1985).

[36] Yew, T. R., and R. Reif, *J. Appl. Phys.*, 68: 4681 (1990).

[37] Hsieh, J. J., D. E. Ibbotson, J. I. Mucha, and D. L. Flamm, In *1989 Proc. Sixth International IEEE VLSI Multilevel Interconnection Conference*, Santa Clara, CA, 12–13 June 1989 (IEEE, 1989), p. 411.

[38] Hsieh, J. J., D. E. Ibbotson, and J. I. Mucha, *Mat. Res. Soc. Symp. Proc.*, 165: 79 (1990).

[39] Einspruch, N. G., and D. M. Brown, eds., *VLSI Electronics Microstructure Science*. New York: Academic Press, 1984.

[40] Chittick, R. C., J. H. Alexander, and H. F. Sterling, *J. Electrochem. Soc.*, 116: 77 (1966).

[41] Spear, W. E., and P. G. LeComber, *Solid State Commun.*, 17: 1198 (1975).

[42] Carlson, D. E., and C. R. Wronski, *Appl. Phys. Lett.*, 28: 672 (1976).

[43] Kuwano, Y., and M. Ohnishi, *J. Phys.* (France), C4, 42: 1155 (1980).

[44] Veprek, S., and M. Heintze, *Plasma Chem. Plasma Process.*, 10: 3 (90).

[45] Gallagher, A., *J. Appl. Phys.*, 64: 2406 (1988).

[46] Kushner, M. J., *J. Appl. Phys.*, 63: 2532 (1968).

[47] Kanicki, J., E. Hasan, D. F. Kotecki, T. Takamori, and J. A. Griffith, In *Amorphous Silicon Technology Symposium*, p. 173. Pittsburgh, PA: MRS, 1989.

[48] Kanicki, J., F. R. Libsch, J. Griffith, and R. Polastre, *J. Appl. Phys.*, 69: 2339 (1991).

[49] Madan, A., and M. P. Shaw, *The Physics and Applications of Amorphous Semiconductors*. San Diego, CA: Academic Press, 1988.

[50] Kuwano, Y., In *Plasma Deposited Thin Films*, eds. J. Mort and F. Jansen, p. 161. Boca Raton, FL: CRC Press, 1986.

[51] Crowley, J. L., *Solid State Technol.*, 35: 94 (1992).

[52] Uchida, Y., S. Oda, and M. Matsumura, *Electron. Commun. Jpn. 2, Electron.*, 72: 48 (1989).

[53] Sassaki, C. A., A. T. Arasaki, M. P. Carreno, A. Komazawa, and I. Pereyra, *J. Non-Cryst. Solids*, 115, 90 (1989).

[54] Kanicki, J., and N. Yoke, *Mat. Res. Soc. Symp. Proc.*, 68: 167 (1986).

[55] Hirose, M., In *Plasma Deposited Thin Films*, eds. J. Mort and F. Jansen, p. 21. Boca Raton, FL: CRC Press, 1986.

[56] Sherman, A., *Thin Solid Films*, 113: 135 (1984).

[57] Smith, D. L., *Mat. Res. Soc. Symp. Proc.*, 165: 69 (1990).

[58] Chow, R., W. A. Lanford, and R. S. Rossler, *J. Appl. Phys.*, 53: 5630 (1982).

[59] Kanicki, J., *IBM Report RC 13750*. Yorktown, NY: IBM Research Division, 1988.

[60] Adams, A. C., In *Plasma Deposited Thin Films*, eds. J. Mort and F. Jansen, p. 129. Boca Raton, FL: CRC Press, 1986.

[61] Kiermaz, A., and K. Beekman, *Semicond. Int.*, 108 (June 1990).

[62] Quinn, W. E., B. G. Bagley, B. J. Wilkens, B. M. Gallois, and D. L. Flamm, *Mat. Res. Soc. Symp. Proc.*, 165: 107 (1990).

[63] Joyce, R. J., H. F. Sterling, and J. H. Alexander, *Thin Solid Films*, 1: 481 (1967).

[64] Lucovski, J., D. V. Tsu, and R. J. Markunas, *Mat. Res. Soc. Symp. Proc.*, 68, 323 (1986).

[65] Landheer, D., N. G. Skinner, T. E. Jackman, D. A. Thompson, J. G. Simmons, D. V. Stefanovic, and D. Khatamian, *J. Vac. Sci. Technol.*, A9: 2594 (1991).

[66] Dzioba, S., *Mat. Res. Soc. Symp. Proc.*, 165: 91 (1990).

[67] Nguyen, S. V., D. Dobuzinsky, D. Dopp, R. Gleason, M. Gibson, and S. Friedman, *Thin Solid Films*, 193: 595 (1990).

[68] Ortiz, A., S. Lopez, C. Falcony, M. Farias, L. C. Araiza, and G. Soto, *J. Electron. Mater.*, 19: 1411 (1990).

[69] Chau, T. T., T. V. Herak, D. J. Thomson, S. R. Mejia, D. A. Buchanan, R. D. McLeod, and K. C. Kao, *IEEE Trans. Electr. Insul.*, 25: 593 (1990).

[70] Ray, S. K., C. K. Maiti, S. K. Lahiri, and N. B. Chakraborti, *Semicond. Sci. Technol.*, 5: 361 (1990).

[71] Nguyen, A. M., and S. P. Murarka, *J. Vac. Sci. Technol.*, B8: 533 (1990).

[72] Emesh, I. T., G. D'Asti, J. S. Mercier, and P. P.Leung, *J. Electrochem. Soc.*, 136: 3404 (1989).

[73] Ibbotson, D. E., J. J. Hsieh, D. L. Flamm, and J. I. Mucha, In *Monitoring and Control of Plasma Enhanced Processing of Semiconductors*, Vol. 1037, p. 130. Santa Clara, CA: SPIE, 1989.

[74] Sterling, H. F., and R. C. G. Swann, *Solid State Electron.*, 8: 653 (1965).

[75] Rahman, M. M., C. Y. Yang, D. Sugiarto, A. S. Byrne, M. Ju, K. Tran, K. H. Lui, T. Asano, and W. F. Stickle, *J. Appl. Phys.*, 67: 7065 (1990).

[76] Hamakawa, Y., Y. Matsumoto, G. Hirata, and H. Okamoto, In *Materials Issues in Microcrystalline Semiconductors Symposium*, Boston, MA, 29 Nov.–1 Dec., 1989 (1990), p. 291.

[77] Fukazawa, T., K. Sasaki, and S. Furakawa, In *Amorphous and Crystalline Silicon Carbide II. Recent Developments. Proc. 2nd. Int. Conf.*, Santa Clara, CA, 15-16 Dec. 1988, p. 49. Berlin: Springer-Verlag, 1989.

[78] Mort, J., and F. Jansen, eds., *Plasma Deposited Thin Films*. Boca Raton, FL: CRC Press, 1986.

[79] Kuhman, D., S. Gramatica, and F. Jansen, *Thin Solid Films*, 177: 253 (1989.).

[80] Carbone, A., F. Demichelis, G. Kanadiakis, G. D. Mea, F. Freire, and P. Rava, *J. Mater. Res.*, 5: 2877 (1990).

[81] De Gangully, G., S. C. Ray, and A. K. Barua, *J. Appl. Phys.*, 69: 3915 (1991).

[82] Matsumoto, Y., G. Hirata, H. Takakura, H. Okamoto, and Y. Hamakawa, *J. Appl. Phys.*, 67: 6358 (1990).

[83] Chang, L. S., P. L. Gendler, and J. H. Lou, *J. Mater. Sci.*, 26: 1882 (1991).

[84] Angus, J. C., P. Koidl, and S. Domitz, In *Plasma Deposited Thin Films*, eds. J. Mort and F. Jansen, p. 89. Boca Raton, FL: CRC Press, 1986.

[85] Grill, A., and B. S. Meyerson, In *Synthetic Diamond: Emerging CVD Science and Technology*, eds. K. E. Spear and J. P. Dismukes, New York: John Wiley & Sons, 1994.

[86] Feldman, A., and S. Holly, *Diamond Optics, Proc. SPIE*. Bellingham, WA: SPIE, 1989.

[87] Clausing, R. E., L. L. Horton, J. C. Angus, and P. Koidl, *Diamond and Diamond-like Films and Coatings, NATO-ASI Series B: Physics*, Vol. 266. New York: Plenum Publishing, 1991.

[88] Aisenberg, S., and R. Chabot, *J. Appl. Phys.*, 42: 2953 (1971).

[89] Tsai, H., and D. B. Bogy, *J. Vac. Sci. Technol.*, A5: 3287 (1987).

[90] Vandentop, G. J., M. Kawasaki, R. M. Nix, I. G. Brown, M. Salmeron, and G. A. Samorjai, *Phys. Rev.*, B41: 3200 (1990).

[91] Catherine, Y., In *Diamond and Diamond-like Films and Coatings, NATO-ASI Series B: Physics*, eds. R. E. Clausing, L. L. Horton, J. C. Angus, and P. Koidl, Vol. 266, p. 193. New York: Plenum Publishing, 1991.

[92] Koidl, P., C. Wild, R. Locher, and R. E. Sah, In *Diamond and Diamond-like Films and Coatings, NATO-ASI Series B: Physics*, eds. R. E. Clausing, L. L. Horton, J. C. Angus, and P. Koidl, Vol. 266, p. 243. New York: Plenum Publishing, 1991.

[93] Robertson, J., *Adv. Phys.*, 35: 317 (1986).

[94] Bubenzer, A., B. Dischler, G. Brandt, and P. Koidl, *J. Appl. Phys.*, 54: 4590 (1983).

[95] Grill, A., B. Meyerson, V. Patel, J. A. Reimer, and M. A. Petrich, *J. Appl. Phys.*, 61: 2874 (1987).

[96] Grill, A., V. Patel, and B. S. Meyerson, In *Diamond and Diamond-like Films and Coatings, NATO-ASI Series B: Physics*, eds. R. E. Clausing, L. L. Horton, J. C. Angus, and P. Koidl, Vol. 266, p. 417. New York: Plenum Publishing, 1991.

[97] Tzeng, Y., M. Yoshikawa, M. Murakawa, and A. Feldman, eds., *Applications of Diamond Films and Related Materials, Materials Science Monographs*. New York: Elsevier, 1991.

[98] Angus, J. C., and C. C. Hayman, *Science*, 241: 913 (1988).

[99] Bachmann, P., *Phys. World*, 4: 32 (1991).

[100] Kamo, M., Y. Sato, S. Matsumoto, and N. Setaka, *J. Cryst. Growth*, 62: 642 (1983).

[101] Badzian, A. R., and R. C. DeVries, *Mater. Res. Bull.*, 23: 385 (1988).

[102] Wei, J., H. Kawarada, J. Suzuki, and H. Hiraki, *J. Cryst. Growth*, 99, 1201 (1990).

[103] ASTeX, *Applied Science and Technology*, Inc., 35 Cabot Road, Woburn, MA 01801.

[104] Aklufi, M., and D. Brock, In *Proc. Second Int. Symp. Diamond Materials*, eds. A. J. Purdue, J. C. Angus, R. F. Davis, B. M. Meyerson, K. E. Spear, and M. Yoder, p. 39. Pennington, NJ: Electrochemical Society, 1991.

[105] Harris, S. J., *J. Mater. Res.*, 5: 2313 (1990).

[106] Frenklach, M., and K. E. Spear, *J. Mater. Res.*, 3: 133 (1988).

[107] Li, Y., and F. E. Talke, *Trans. ASME*, 112: 670-7 (1990).

[108] Harker, A. B., and J. F. DeNatale, *J. Mater. Res.*, 5: 818 (1990).

[109] Harris, S. J., and A. M. Weiner, *Appl. Phys. Lett.*, 55: 2179 (1989).

[110] Liou, Y., A. Inspector, R. Weimer, D. Knight, and R. Messier, *J. Mater. Res.*, 5: 2305 (1990).

[111] Kamo, M., and Y. Sato, In *Proc. Second Int. Symp. Diamond Materials*, eds. A. J. Purdue, J. C. Angus, R. F. Davis, B. M. Meyerson, K. E. Spear, and M. Yoder, p. 20. Pennington, NJ: Electrochemical Society, 1991.

[112] Shing, Y. H., F. S. Pool, and D. H. Rich, In *Applications of Diamond Films and Related Materials, Materials Science Monographs*, eds. Y. Tzeng, M. Yoshikawa, M. Murakawa, and A. Feldman, Vol. 73, p. 497. New York: Elsevier, 1991.

[113] Suzuki, J., H. Kawarada, K. Mar, J. Wei, Y. Yokota, and A. Hiraki, *Jpn. J. Appl. Phys.*, 28: L281 (1989).

[114] Baik, Y., and K. Eun, In *Applications of Diamond Films and Related Materials, Materials Science Monographs*, eds. Y. Tzeng, M. Yoshikawa, M. Murakawa, and A. Feldman, Vol. 73, p. 521. New York: Elsevier, 1991.

8

Plasma Assisted Etching

8.1 INTRODUCTION

Etching is used in microelectronic processing for pattern transfer during fabrication of devices. The pattern to be transferred is defined by a mask prepared by lithographic techniques on layers of photoresist. The photoresist materials, which are used to delineate the patterns to be etched, often lose adhesion in the solutions used for wet etching processes and thereby alter pattern dimensions and prevent control of line width. When wet etching progresses downward into the film, it also proceeds laterally at approximately equal rate, as illustrated in Fig. 3-13 (a) in Chapter 3. This results in an isotropically etched profile and causes undercutting under the mask as shown in Fig. 8-1 (a). The described type of etching, which proceeds at similar speeds in all directions is called *isotropic etching*. When the thickness of the etched film is small relative to the minimum pattern dimension the undercutting is not significant. However, when the film thickness is comparable with pattern size, as is the case for current and future devices, undercutting becomes intolerable. In addition, as device geometries decrease, spacing between resist stripes decreases, making undercutting prohibitive.

Problems encountered while using wet etching in solutions can be overcome by *plasma* or *dry etching*. Plasma etching is performed using a discharge in a reactive gas to create reactive atoms and radical species that react with surfaces to form volatile compounds. The formed products evaporate from the surface leaving an etched substrate. When using plasma etching, adhesion is no longer a major problem, undercutting can be controlled by varying the plasma chemistry and parameters, and directional or anisotropic profiles can thus be generated.

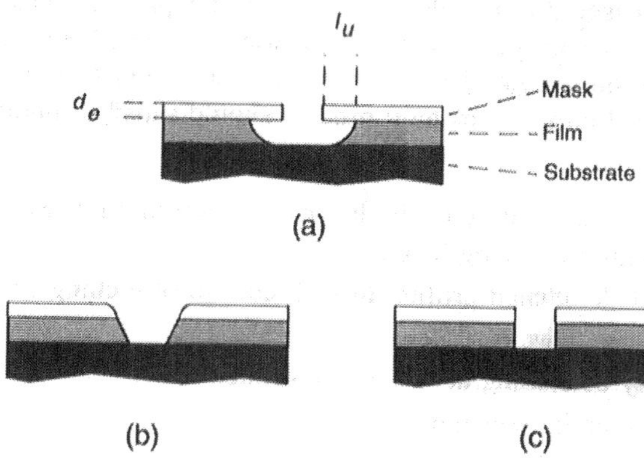

Fig. 8-1 Types of plasma etching: (a) isotropic etch; (b) sloped wall via etch; (c) vertical wall (anisotropic etch).

Detailed discussions of plasma etching and the mechanisms of plasma etching of different materials can be found in several monographs and review papers, such as references [1–6].

Etch directionality is required in the fabrication of features of high-aspect ratios such as the deep trenches (> 5 μm) used for storage capacitors in silicon wafers. The etching process is called *anisotropic etching* when it proceeds much faster in one direction as compared to other directions. The directionality, or anisotropy, is defined by the ratio of vertical to horizontal etching rate. The *degree of isotropic etching, IE*, can be defined by [7]

$$IE = \frac{l_u}{d_e} \tag{8.1}$$

where l_u = amount of undercut at end of etching (see Fig. 7-1)
d_e = thickness of etched layer

The *degree of anisotropic etching* is correspondingly

$$AE = 1 - IE \tag{8.2}$$

In addition to isotropy / anisotropy, another important etching characteristic is the *etch selectivity*. Etch selectivity between two materials is defined as the ratio between their etching rates at identical plasma conditions. Etch selectivity usually refers to a high *etching rate ratio* between chemically different materials or between the etched and the underlying layer. Etch selectivity is required between etch mask and substrate or between different substrate materials.

Plasma etching was first used by Irving [8] and has since evolved and advanced to the stage that makes it possible to produce the very dense VLSI chips with characteristic dimensions smaller than 1 μm. The plasma etching has developed into several different processes, such as *plasma etching, reactive ion etching* (*RIE*), and more recent *magnetically enhanced reactive ion etching* (*MERIE*) and *ECR etching.*

The main reason for the development of the plasma enhanced chemical processing techniques and expansion of their use over wet chemical etching is the ability to achieve good control of the processes necessary to obtain the required shape of the etched area. An etching process should satisfy a number of requirements, including:

- High etching selectivity of the layer to be etched with respect to the mask material and underlying layers
- Control of the etched profile through control of etching anisotropy
- High etching rates
- Uniformity of etching across wafer surface
- Minimal material damage
- Residue-free surface after etching.

The processes taking place during plasma etching can be grouped into four categories as illustrated in Fig. 8-2 [2]:

1. Sputtering
2. Chemical volatilization
3. Ion enhanced chemistry
4. Inhibitor ion enhanced chemistry

The mechanisms of these processes can be separated into chemical and physical etching. Chemical volatilization, or chemical etching, illustrated in Fig. 7-2(b), is caused by neutral species generated in the plasma, which react with the etched layer to form volatile compounds. Chemical etching is typically isotropic and is characterized by high etch rates, usually good etch selectivity, and low substrate damage. Sputtering, Fig. 8-2(a), is a purely physical process caused by surface bombardment with high-energy ions. It can provide anisotropic etching, but it is also associated with poor selectivity, high surface damage, and low etching rates.

Ion enhanced chemistry, Fig. 8-2(c), is a combination of physical and chemical processes in which the chemical reactions at the surface of the processed sample are enhanced by ion bombardment, as described in Sec. 3.5. The fourth process, shown in Fig. 8-2(d), is an ion enhanced process but that also involves an inhibitor that passivates the surfaces that are not exposed to ion bombardment. This process is exploited in sidewall passivation for anisotropic etching (see Sec. 8.3.2). By choosing the suitable reactor and adjusting the plasma parameters, it is possible to control the interaction between the physical and chemical mechanisms to achieve the required etching characteristics.

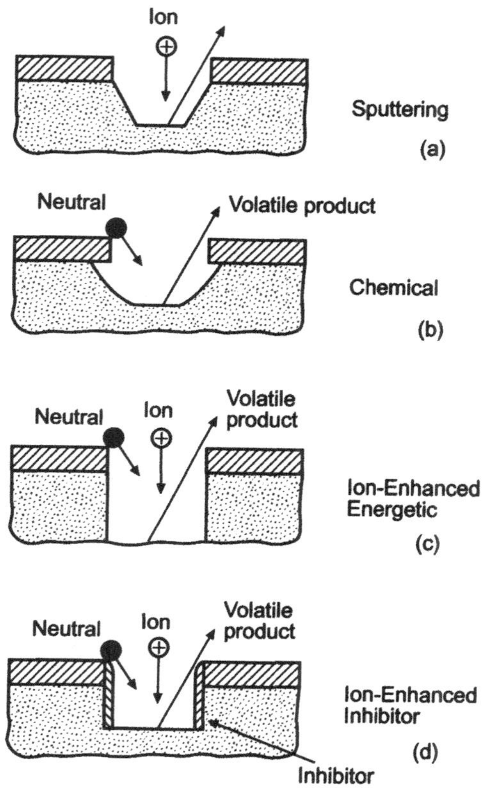

Fig. 8-2 Processes taking place during plasma etching (from [2], reprinted with permission from *Plasma Etching*, p. 93, 1989).

8.2 PLASMA ETCHING

Plasma etching generally refers to etching performed under low ion bombardment and caused mainly by chemical reactions with plasma species. Plasma etching is performed under conditions at which the physical effects are negligible and is usually an isotropic etching process. The role of the plasma in this case is only to produce the chemically active species. The process is therefore similar to a large extent to wet etching with the difference that the process occurs in the gas phase. Plasma etching is performed at relatively high pressures, larger than 0.1 torr, often about 1 torr. If a parallel plate reactor is used, the substrates are placed on the grounded electrode as shown in Fig. 8-3. Under these conditions the substrates are at a low potential relative to the plasma and ion bombardment of the substrates is very weak.

Plasma etching is also performed in barrel reactors, such as shown in Fig. 4-4 in Chapter 4, often with etch tunnel, whose role is to limit the discharge to a

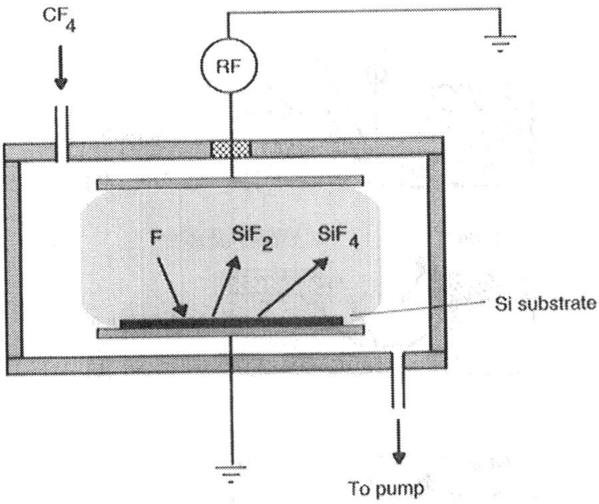

Fig. 8-3 Plasma etching in a parallel plate reactor.

region outside the substrates. The substrates interact only with plasma species diffusing out of the plasma. Another possibility is etching in the plasma effluents, downstream of the plasma, in a remote plasma reactor, as illustrated in Fig. 8-4. In this arrangement, only neutrals reach the wafer, which is chemically etched by atoms and radicals.

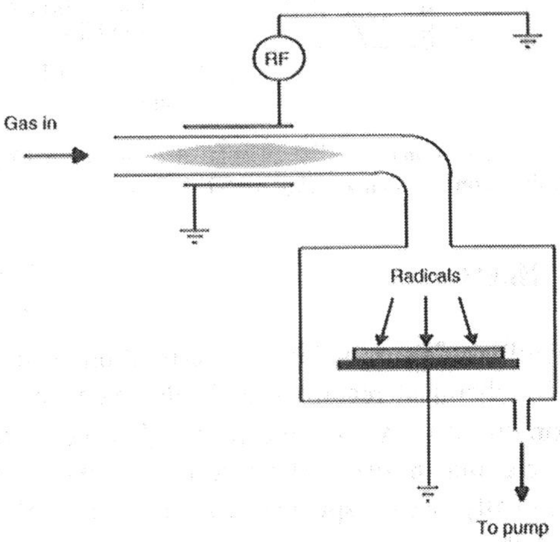

Fig. 8-4 Plasma etching in a downstream configuration.

8.2.1 Mechanisms of Plasma Etching

Plasma etching, illustrated in Fig. 8-3, is similar to plasma ashing and proceeds through the following steps:

1. Active species are formed in the plasma from non-reactive gases.

2. The active species formed in the plasma adsorb on the surface of the substrate and react with the surface to form volatile products.
3. The volatile products leave the treated surface and are pumped out with the effluents of the plasma; these formed products should have a high vapor pressure at the substrate temperature, usually below 100 °C.

The effect of plasma gas composition on the etching process has been explained by the "etchant-unsaturated species" model developed by Flamm [7, 9]. The model is based on the different possible reaction paths in the plasma, initiated by electron impact and is described by the following equations:

$$e + \text{halocarbon} \rightarrow \begin{matrix}\text{saturated}\\ \text{species}\end{matrix} + \begin{matrix}\text{unsaturated}\\ \text{species}\end{matrix} + \text{atoms} \quad (8.3)$$

$$\begin{matrix}\text{reactive atoms,}\\ \text{reactive molecules}\end{matrix} + \begin{matrix}\text{unsaturated}\\ \text{species}\end{matrix} \rightarrow \begin{matrix}\text{saturated}\\ \text{species}\end{matrix} \quad (8.4)$$

$$\text{atoms} + \text{surfaces} \rightarrow \text{chemisorbed layer} + \text{volatile products} \quad (8.5)$$

$$\text{unsaturated species} + \text{surfaces} \rightarrow \text{films} \quad (8.6)$$

The corresponding reactions in a CF_4 plasma are

$$2e + 2CF_4 \rightarrow CF_3 + CF_2 + 3F + 2e \quad (8.7)$$

$$F + CF_2 \rightarrow CF_3 \quad (8.8)$$

$$4F + Si \rightarrow SiF_4 \quad (8.9)$$

$$nCF_2 + \text{surface} \rightarrow (CF_2)_n \quad (8.10)$$

According to this model, the final result of the interaction between a CF_4 plasma and silicon depends on the generated species. The result will be etching, if the main path in the plasma reactions follows Eq. (8.9); recombination, if the main path is described by Eq. (8.8); or film formation (polymerization), if the main path is described by Eq. (8.10). Gas-phase oxidant additives (O_2, F_2 etc.) can dissociate and react with unsaturated species, changing the relative concentration of the species and the reaction paths.

8.2.2 Loading Effects

Most isotropic etchants exhibit a *loading effect*. The loading effect is the result of a high consumption rate of the etchant by the etched film, causing the depletion of the active etchant consumed during the etching process. If the etchant is depleted mainly by surface reactions, then small increases in flow rates will cause large increases in etching rates. When loading occurs, the overall etching rate depends upon the total area of film to be etched. As the process approaches the "end point" of the etching, the area of the etched material decreases, the concentration of the active species increases, and a drastic increase in the etching rate occurs. As the etching is isotropic, this causes a rapid increase in the lateral etching [7], and severe undercutting can occur if the process is not stopped in time. Under extreme loading conditions, when carbon-containing gases are used, etchant depletion can be so severe that instead of etching, polymer deposition takes place.

Analysis of the loading effect indicates that the etching rate for a batch of N wafers, each of area A, is given by [10]

$$R(N) = \frac{(k_{etch} / k_{loss})G}{1 + (k_{etch}\rho NA / k_{loss} Vol)} \tag{8.11}$$

where R = overall etching rate
k_{etch} = etching rate constant at the limit of an "empty" reactor
k_{loss} = etchant loss at the limit of an "empty" reactor
ρ = the number density of substrate molecules per unit wafer area in the reactor
G = generation rate of the active etchant species
Vol = volume of the reactor.

Equation (8.11) indicates that when the total wafer area is large enough so that

$$\frac{k_{etch}\rho NA}{k_{loss} Vol} \gg 1 \tag{8.12}$$

the etching rate can be expressed as

$$R(N) = \frac{GVol}{\rho NA} \tag{8.13}$$

that is, in these conditions the etching rate varies inversely with the number of wafers. This dependence is typical of a strong loading effect and indicates that loading effects can be reduced to some extent by using large-volume reactors. The loading effect is eliminated when etchant loss, either homogeneous or heterogeneous, dominates ($k_{loss} \gg k_{etch}$). In such a case the etching rate is no longer a function of the total wafer area NA.

Transport limitations, such as diffusion controlled reactions, can lead to localized depletion of etchant (i.e., local loading), causing etching nonuniformities. The size and density of features on the substrate can influence the etching rate at different locations on a single wafer and thus produce pattern sensitivity of etching. Depletion of the etchant gases across a wafer produces a bull's-eye effect, while depletion across a reactor is indicated by the fact that the leading wafer edge etches faster than the trailing edge.

Similar effects are observed when the removal of the reaction products is transport limited. The effect depends on the correlation of the residence time, diffusion, convection, and reaction rates. If the residence time in the reactor is smaller than reaction time, the influence of the diffusion is reduced, while long residence times will enhance them [3].

Most of the described effects can be reduced by a correct choice of pressure and flow rates. One approach to eliminate the loading problem in manufacturing reactors is to operate them under high flow rates, thus increasing the concentration of the etchant. This way the changes in the surface reactions near the end of the etching do not significantly affect the etching rate [11].

8.3 REACTIVE ION ETCHING

Reactive ion etching is the most used dry etching technique. RIE is based on a combination of chemical activity of reactive species generated in the plasma with physical effects caused by ion bombardment. When substrates to be etched are positioned on the powered electrode of a parallel plate reactor, as illustrated in Fig. 8-5, a reactive ion etching or reactive sputter etching configuration results. A negative DC bias is generated on the sample electrode, usually by applying RF power to the electrode. The electrode and the samples placed on it are therefore exposed during RIE to the bombardment with energetic positive ions, having energies between 300 and 700 V.

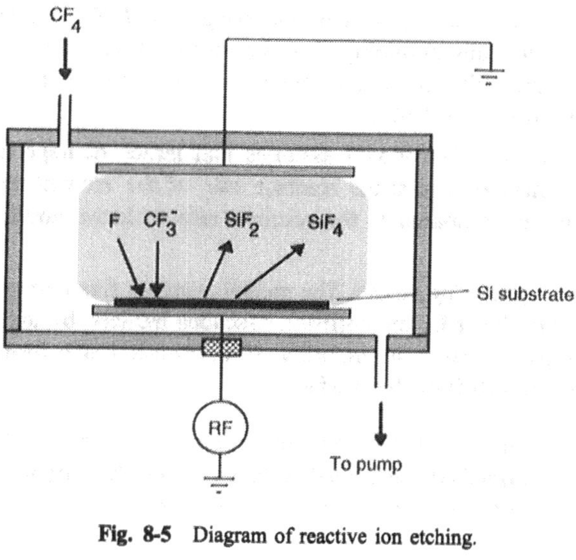

Fig. 8-5 Diagram of reactive ion etching.

The RIE arrangement generally causes the wafers to be subjected to higher-energy ion bombardment than in the plasma etching mode because of the larger negative potentials established on the powered electrode as compared to the grounded electrode. Another reason for higher-energy ions in RIE is the lower operating pressure of 10–200 mtorr. No exact pressure demarcation exists, but RIE is typically carried out at pressures below 13 Pa (0.1 torr), whereas plasma etching is performed at higher pressures, which cause an additional reduction of the average energy of the ions bombarding the surface.

8.3.1 Mechanisms of RIE

In an RIE process the surface is exposed simultaneously to both energetic positive ions and thermal radicals or atoms. The chemical effect of the radicals is the same as in plasma etching (see Sec. 8.2.1). The ion bombardment strongly enhances the process, in some instances up to two orders of magnitude. The increase is much larger than that expected from the sputtering of the surface by ions. This is the so called ion assisted gas-surface chemistry or chemical sputtering, described in Sec. 3.5.

The microscopic details of the ion assisted chemical process are not well understood, but ion induced nuclear motion is believed to be involved because electrons at the same energy are two orders of magnitude less efficient in this process [12].

Several mechanisms have been proposed to explain the *ion enhanced etching* [13]:

1. *Chemically enhanced physical sputtering model.* The model explains the effect of the ions assuming that the surface layer, which is chemically modified through ion bombardment, has a larger sputtering yield than the unmodified surface [14].
2. *Damage model.* The model assumes that lattice damage induced by ion bombardment enhances the reaction rate of the etchant species with the substrate, as compared to the reaction rate with the undamaged material [7].
3. *Chemical sputtering model.* The model assumes that energy is supplied to the reaction layer by the collision cascades induced by ion bombardment; this energy increases the mobility of molecules which form volatile products and desorb from the surface.

According to these models, both the reaction step and the removal of the volatile reaction products formed on the etched surface are enhanced by ion bombardment. The predominance of one mechanism over the other is determined in each case by the etchant-substrate combination. The chemical sputtering model, for example, appears to be accepted as explaining etching in the widely used fluorine silicon system [15].

Most of the materials used in VLSI processing can be plasma etched or reactively ion etched. They react with species formed in plasmas, usually halogen atoms, to form volatile compounds. Si, SiO_2, Si_3N_4, GaAs, polyimide, photoresist, Al, W, Mo, and Ti, can be dry etched in a plasma.

The halogenation of materials in an etching process is believed to occur through adsorption of the halogen atom on the surface and abstraction of an electron from the solid to become a negative ion. To obtain full halogenation for forming volatile molecules such as SiF_4, some penetration of the halogen into the crystal lattice has to occur. The penetration is thought to be assisted by the electric field and depends on the size of the etchant species. This is indicated by the fact that while the small H and F atoms etch silicon effectively, the larger Cl atoms etch it slowly, and the even larger Br atoms do not etch silicon at all at room temperature, unless the surface is bombarded simultaneously with energetic particles.

While most reaction rates of reactive species with surface are enhanced by bombardment with energetic particles, there are some exceptions, for example, silicon etching by hydrogen and aluminum and copper etching by chlorine [12].

8.3.2 Anisotropy

Optimization of plasma etching for tight profile control is very important when considering that in present VLSI technology one has to etch very narrow (less than 1 μm wide) and deep (10 μm or more) profiles, uniformly over the whole area of a 8-inch-diameter silicon wafer. The sidewalls of the trench must be perfectly straight, as illustrated in Fig. 8-1 (c). To produce this type of trench, the etching has to be directional.

Etch directionality is generally achieved by input of directed energy on the surface exposed to a chemical etchant. The energy input can by supplied by ion, electron, or photon bombardment. In RIE, etch directionality is attained by bombardment with heavy energetic ions, such as CF_3^+. Ion assisted chemistry causes the ion bombarded bottom of the etched feature to etch much faster than the sidewalls, which are not bombarded by ions and thus directionality is obtained.

The effect that the etching rate dependence on ion bombardment has on the shape of the etched profile is illustrated in Fig. 8-6. The figure is representative for a plasma gas that etches silicon at a rate $V_x > 0$, even in the absence of ion bombardment, but does not etch silicon dioxide without ion bombardment. When a negative bias is applied to the substrate, silicon dioxide will be etched only in the direction normal to the surface (direction of ion bombardment), resulting in an etched profile with vertical walls. Silicon will be etched in the same plasma in both directions, normal and parallel to the surface, though faster in the normal direction. The profile etched in silicon will have a sloped profile, with the slope depending on the ratio of the etch rates in the two directions (see Fig. 8-6):

$$\frac{X}{Z} = \frac{V_x}{V_z} \tag{8.14}$$

where V_x = etching rate at zero bias

V_z = etching rate at applied negative bias.

The chemistry of the plasma has a very important role in the etching process. Etching in SF_6 plasma can provide high rates of isotropic etching, while Cl_2 plasma yields lower etching rates but provides a much better profile control. By adjusting the plasma chemistry and its parameters, it is possible to change the etch directionality between isotropic and completely anisotropic. It is thus possible to adjust the plasma parameters to etch sidewalls of required slopes, such as those required for interconnect vias in VLSI shown in Fig. 8-1 (b).

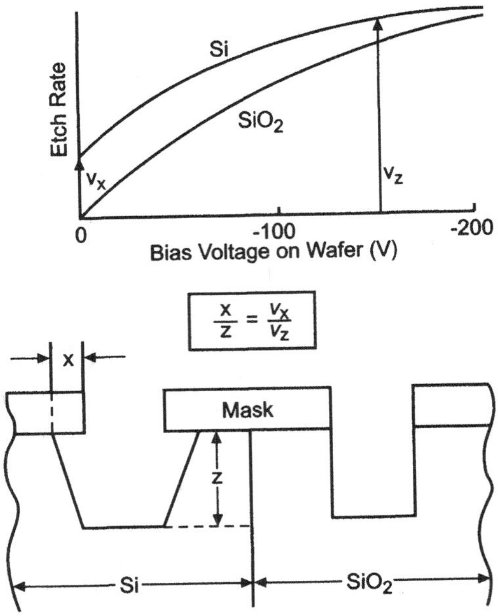

Fig. 8-6 Shape of etched profile as a function of etching rate dependence on substrate bias (from [16], reprinted with permission from S. Veprek and M. Venugopalan, eds., *Topics in Current Chemistry, Plasma Chemistry III,* Berlin: Springer-Verlag, 1980).

8.3.2.1 Sidewall Passivation

Undoped silicon can be etched in chlorine or bromine plasmas at room temperature only under ion bombardment, and vertical profiles can be reactively ion etched in those plasmas. By comparison, highly doped *n*-type silicon etches easily in a chlorine plasma, even without ion bombardment [17]. In general, *n*-type silicon etches faster than intrinsic silicon, which in turn etches faster than *p*-type silicon [18]. The dopant concentration has to be higher than 10^{19} to observe the doping effect on the etching rate.

During manufacturing of VLSI, etching has to be done often through several layers of different types or degrees of doping. Because of the different lateral etching rates of these layers, the control of the etching profile by pure ion bombardment is difficult. This problem is solved by the *sidewall passivation*, or *sidewall blocking*, effect [7, 19].

The sidewall passivation is a result of the formation on the sidewall of the etched feature of a film, which then slows down or completely stops lateral etching. For example, by adding oxygen to a chlorine plasma, a film can grow on the sidewall of the feature that is not exposed to ion bombardment, while the growth is prevented at the bottom of the feature by ion bombardment. The etching continues therefore only at the bottom of the feature, resulting in formation of vertical walls.

The formation of the passivating film is illustrated in Fig. 8-7 for etching of silicon with a chlorine-containing plasma. The sidewall film was found, in this

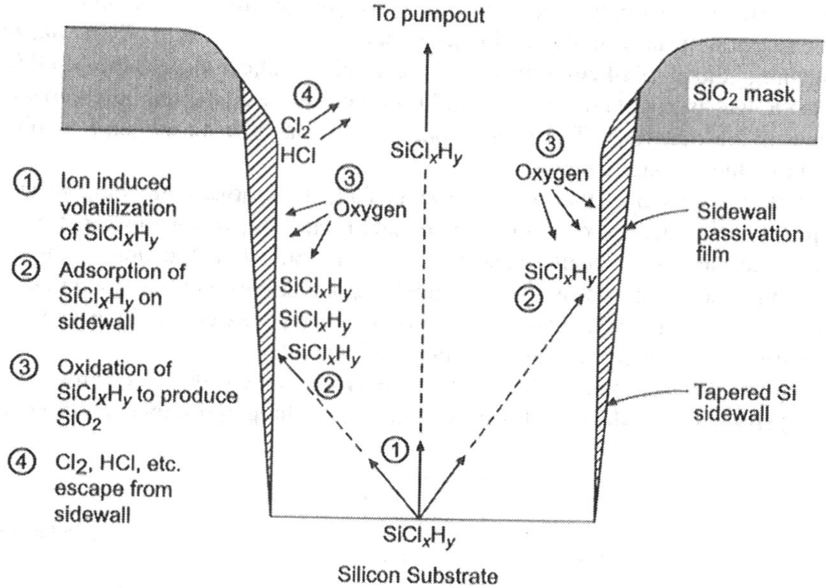

Fig. 8-7 A model for sidewall film formation in a $HCl/O_2/BCl_3$ RIE (from [20], reprinted with permission).

case, to be nearly stoichiometric SiO_2 [20]. The sidewall layer grows from the radicals formed by the etching of the silicon surface at the bottom of the trench. These radicals, $SiCl_4$, $SiCl_3$, $SiCl_2$, SiCl, and Si, reach the sidewalls where they adsorb with a large sticking coefficient. There, they react with oxygen from the plasma to form SiO_2 and volatile Cl_2, HCl, and Cl_2O, which desorb from the surface. The formed silicon dioxide film has a low etching rate and prevents the lateral etching of the walls.

Sidewall blocking can be achieved even in the absence of oxygen. As the stoichiometry of the feed gas is made halogen deficient, a condition can be reached in which carbon-containing species polymerize and form a protective film that stops the etching of the substrate almost completely. The etching rate of the carbon-containing species is strongly enhanced by the energetic ion bombardment. The ion bombardment on the bottom surface prevents the formation of the carbonaceous polymer film, and the downward etching continues unaffected, while the lateral etching is stopped.

Using the sidewall passivation approach, the etch profile can be controlled satisfactorily; however, an additional cleaning step is required at the end of the etching process to remove the sidewall film, before further processing.

8.3.3 Selectivity

Etch selectivity is caused by the different rates of the processes taking place on the different materials. These processes include adsorption of etching species and desorption of the volatile etching products. Using $CF_4 + H_2$ or CHF_3 plasmas for etching silicon or silicon dioxide, it is possible to adjust the plasma conditions in such a way as to obtain almost equilibrium between etching and polymerization of one of the materials. The plasma parameters can then be adjusted to obtain etching of the second material.

It was shown by Coburn and Winters that, in fluorine-containing plasmas, the process can change from etching to polymerization, depending on the F/C ratio in the gas phase etching species [21]. This ratio is a function of the F/C ratio in the plasma gas but is also affected by gas additives and plasma conditions, as illustrated in Fig. 8-8, which shows the boundary between etching and polymerization as a function of substrate bias and gas additives.

Figure 8-8 shows that the substrate bias can affect the F/C ratio in the etching species and affect the boundary between etching and polymerization. The

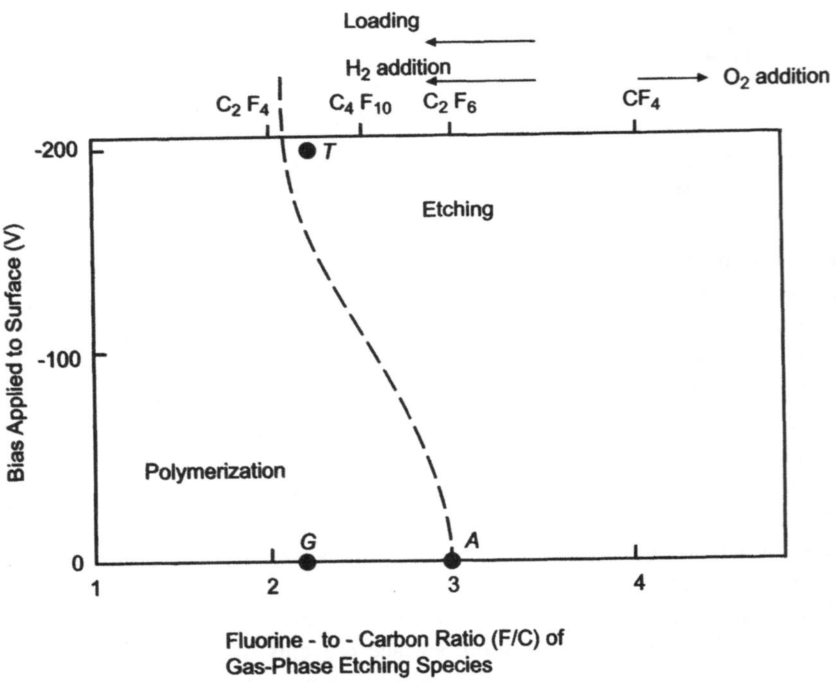

Fig. 8-8 Effect of plasma composition and bias on the etching/polymerization boundary (from [21], reprinted with permission).

described effect is exploited to achieve etching selectivity. The etchant unsaturated model of Flamm and Donnelly [7, 9] and the fluorine-to-carbon ratio concept of Coburn and Winters [21] are similar because both models are based on stoichiometric arguments [11]. In both models it is assumed that the etching is produced mainly by atomic radicals and that gas mixtures that increase the concentration of these radicals in the plasma increase the etching rates.

In fluorocarbon plasmas the main etching species are F atoms. These can be lost through gas phase recombination reactions with CF_3 radicals. In addition, CF_x species can form fluorocarbon films on the treated surfaces and block the etching of the silicon. If the feed gas contains additives that reduce the concentration of unsaturated species or increase the ratio of active fluorine to carbon, the etching rate will increase according to the two models. Thus, in a CF_4 plasma, addition of oxygen, which reacts with the carbon atoms to form CO, CO_2, COF, and COF_2 increases the F/C ratio in the etching species. The effect of hydrogen addition is opposite, because hydrogen scavenges fluorine atoms and decreases the etching rate. Hydrogen is therefore used to reach selective etching of silicon dioxide over silicon. In such a case, the Si surface is blocked by adsorbed fluorocarbon and relatively few F atoms reach the Si and etch it. On the silicon dioxide surface, there is a supply of oxygen atoms from the substrate to react with adsorbed carbon and form volatile CO and CO_2 and the fluorine reacts with silicon to form the volatile CF_4.

Similarly, in CHF_3 plasmas, the hydrogen produced by the dissociation of CHF_3 reduces the Si surface exposed to the fluorine plasma and initiates the polymerization of CF_3 radicals. On the SiO_2 surface, the oxygen liberated by the RIE process reacts with the fluorocarbon radicals to form volatile carbon compounds and the etching continues unimpeded [22].

Using such plasmas with silicon and silicon dioxide, fluorocarbon film deposits on the silicon surface preventing complete fluorination of the silicon and reducing its etching. At the same time, the surface of the silicon dioxide remains free of the polymer film and etching takes pace at a faster rate. Etch selectivity of SiO_2 to Si is thus obtained.

Etch selectivity can be achieved also according to the damage model described earlier. Ion bombardment generally imparts different degrees of damage to various materials. By changing the energy of the bombarding ions, etching selectivity can be altered to tailor a specific etching process [3]. Nevertheless, etch selectivity is more difficult to achieve in RIE than in wet chemical etching, because the ion bombardment makes the chemical difference between materials less important.

When ion bombardment controls the reaction, Eq. (8.13) is not applicable, because the etching rate is determined by the bombardment flux rather than the etchant supply. RIE does therefore not have loading effects.

As discussed in Sec. 3.5.4, the bombardment of the substrate with the high-energy ions (up to about 700 eV) can induce damage to the surface layers of the etched wafer. The damaged layer and the fluorocarbon film deposited on the Si surface have to be removed before further processing the wafer [23].

The etching rates are generally low in RIE, and many wafers (6 inches in diameter) have to be processed simultaneously to obtain reasonable commercial throughputs. The transition to processing of larger silicon wafers (presently up to 8 inches in diameter) requires single-wafer processing and, therefore, faster etching rates. Directional faster etching rates (up to ~1 μm/min) were obtained using ECR or magnetically enhanced reactive ion etcher reactors [24]. While etching in RIE is obtained from a small flux of high-energy ions, large fluxes of low-energy ions characterize the ECR and MERIE processes.

A comparison of several etching techniques and their characteristic parameters are presented in Table 8-1.

TABLE 8-1 Comparison of Different Etching Techniques (from [25], reprinted with permission)

Parameters	ICP	ECR	Helium	SPRP	MERIE	RIE	PE
Frequency	13.56 MHz	2.45 GHz	13.56 MHz	400 kHz	13.56 MHz	13.56 MHz	13.56 MHz
Gas pressure (torr)*	$\sim 1^{-3}$	$\sim 4^{-4}$	$\sim 1^{-3}$	$\sim 1^{-1}$	$\sim 1^{-2}$	$\sim 7^{-2}$	$\sim 1^{-1}$
Electron temperature (eV)	~ 4	~ 4	~ 4	~ 10	~ 5	~ 8	~ 8
Plasma density (cm^3)*	$\sim 5^{11}$	$\sim 3^{11}$	$\sim 5^{11}$	$\sim 8^{10}$	$\sim 5^{10}$	$\sim 1^{10}$	$\sim 3^{8}$
Ion current density (mA/cm^2)	~ 10	~ 10	~ 10	~ 2	~ 1	~ 0.1	~ 0.1
Ion energy (eV)	Controllable	Controllable	Controllable	30–150	~ 200	200–1000	~ 20

ICP – inductively coupled plasma
ECR – electron cyclotron resonance
SPRP – split power reverse phase/rainbow 4500
* indicates $\sim \times 10^x$
MERIE – magnetically enhanced RIE
RIE – reactive ion etching
PE – plasma etching

8.4 ETCHING OF SPECIFIC MATERIALS

Most of the materials used in VLSI fabrication, including silicon, silicon dioxide, silicon nitride, gallium arsenide, polyimide, photoresist, aluminum, tungsten, molybdenum, and titanium, can be dry etched by plasma or by reactive ion etching. A diversity of gases are used for plasma etching of different materials, as illustrated in Table 8-2. The etching can be performed in different types of reactors, using various electrode configurations, pressures, and frequencies, as shown in Table 8-3 [26]. As mentioned earlier, plasma etching is generally performed in barrel etchers or downstream reactors. Both types of reactors are characterized by minimal ion bombardment.

TABLE 8-2 Plasma Gases Used for Etching of Different Materials (from [26])

Material	Gases
Silicon	CF_4/O_2, CF_2Cl_2, CF_3Cl, $SF_6/O_2/Cl_2$, $Cl_2/H_2/C_2F_6/CCl_4$, C_2ClF_5/O_2, SiF_4/O_2, NF_3, ClF_3, CCl_4, CCl_3F_5, C_2ClF_5/SF_6, C_2F_6/CF_3Cl, Br_2, CF_3Cl/Br_2
SiO_2	CF_4/H_2, C_2F_6, C_3F_8, CHF_3/O_2
Si_3N_4	$CF_4/O_2/H_2$, C_2F_6, C_3F_8, CHF_3
Organics	O_2, CF_4/O_2, SF_6/O_2
Silicides	CF_4/O_2, NF_3, SF_6/Cl_2, CF_4/Cl_2
Al	BCl_3, BCl_3/Cl_2, $CCl_4/Cl_2/BCl_3$, $SiCl_4/Cl_2$
Cr	Cl_2, CCl_4/Cl_2
Mo, Nb, Ta, Ti, W	CF_4/O_2, SF_6/O_2, NF_3/H_2
Au	$C_2Cl_2F_4$, Cl_2, $CClF_3$
GaAs	BCl_3/Ar, $Cl_2/O_2/H_2$, $CCl_2F_2/O_2/Ar/He$, CCl_4
InP	CH_4/H_2, C_2H_6/H_2, Cl_2/Ar

RIE is performed in parallel plate RF reactors, ECR, or magnetically enhanced reactors. In parallel plate reactors the substrates are usually placed on the RF-powered electrode, while the other electrode is electrically connected to the grounded walls of the reactor. This way the substrate electrode acquires a high negative DC bias. The plasma potential is generally low and prevents contamination by sputtering from the grounded surfaces. A variation of the parallel plate reactor is the hexode reactor, in which the cathode has a hexagonal shape and is surrounded by the round wall of the reactor that forms the anode. The hexode reactor allows processing of larger batches of wafers. Etching of several materials will be discussed next.

8.4.1 Semiconductors

8.4.1.1 Silicon and Polysilicon Etching

Single-crystal silicon and polycrystalline silicon (poly-Si) are etched in halogen-containing plasmas. The etching of silicon in fluorine-containing plasmas is best understood today [27, 28]. The main products in the etching of silicon in fluorine plasmas are SiF_2 and SiF_4. The fluorine atoms chemisorb on the surface of the silicon and, without bombardment with energetic ions, react with the silicon to form a stable, partially reacted layer several atomic layers thick, consisting primarily of SiF, SiF2, and SiF_3 entities [28]. At low pressures (~ 0.013 Pa), a SiF_2-like steady-state surface layer is formed. At higher pressures, SiF_3 radicals are the primary species detected on silicon surfaces exposed to a fluorine-containing plasma [29].

The continuation of the reaction to form the volatile SiF_4 is a relatively slow process and requires that the fluorinated surface layer be penetrated by impinging F

TABLE 8-3 Reactors and Conditions Employed for Plasma Assisted Etching (from [26])

Etch Tool Configuration	Etching Mechanism	Pressure Range (torr)	Electrode Arrangement	Wafer Location and Biasing	Excitation
Barrel etching or tunnel etching	Chemical	$10^{-1} - 10^{0}$	Coil or electrodes outside vessel	In rack in plasma; wafers electrically floating	RF
Downstream or effluent plasma etching	Chemical	$10^{-1} - 10^{0}$	Coil or electrodes outside vessel	In rack downstream of plasma	RF
Plasma etching	Chemical or chemical/physical	$10^{-1} - 10^{1}$	Planar diode	On grounded (anode) electrode	RF
Ion etching or sputter etching	Physical	$10^{-2} - 10^{-1}$	Planar diode or cylindrical (hexagonal) diode	On cathode	RF
Reactive ion etching	Chemical/physical	$10^{-3} - 10^{-2}$	Planar diode or cylindrical (hexagonal) diode	On cathode	RF
Mangetic confinement ion etching	Physical	$10^{-3} - 10^{-2}$	Planar diode or cylindrical (hexagonal) diode with magnetic field confinement near cathode	On cathode	RF
Magnetic confinement reactive ion etching	Chemical/physical	$10^{-3} - 10^{-2}$	Planar diode or cylindrical (hexagonal) diode with magnetic field confinement near cathode	On cathode	RF
Triode etching	Chemical/physical	$10^{-3} - 10^{-2}$	Triode	On platform electrode	RF or DC
Ion beam etching or ion beam milling	Physical	10^{-4}	Planar triode	On grounded (cathode) electrode	DC
Reactive ion beam etching	Chemical/physical	10^{-4}	Planar triode	On grounded (cathode) electrode	DC
Electron cyclotron resonance sources	Chemical/physical	10^{-4}	External	On either	Microwave

atoms. The short-lived but high-energy density collision cascades resulting from ion-solid collisions are believed to accelerate the conversion of this partially reacted layer to SiF_4, most probably by a disproportionation process involving the fluorine. The reaction probability for this step is low (< 0.01) at normal surface temperatures. The penetration is therefore the rate limiting step of the process and, for silicon oriented in the < 100 > direction, it proceeds at a rate defined by [30]:

$$R_F(\mathrm{Si}) = 2.91 \times 10^{-12} T^{1/2} n_F \exp\left(-0.108 \frac{\mathrm{eV}}{kT}\right) \tag{8.15}$$

where $R_F(\mathrm{Si})$ = etching rate of Si in Å/min
n_F = F atom concentration per cm^3
T = the temperature in °K

Most fluorine-containing etchant gases display similar apparent activation energies (~ 0.1 eV per atom), indicating that identical mechanisms and etchant (F atoms) are probably operative in these slightly different chemical systems [3].

Etching of silicon in fluorine-based plasmas results generally in a large undercut under the mask, indicating a large isotropic chemical etching rate.

Fluorine-containing gases such as carbon tetrafluoride (CF_4) and sulphur hexafluoride (SF_6) offer advantages because of their low toxicity. However, these gases can polymerize in the plasma to form unsaturated species such as C_xF_{2x} and S_xF_{2x}, which can scavenge free F atoms from the plasma and, in extreme cases, can lead to significant formation of polymer residue. Oxygen, which is usually added to plasmas of these gases to enhance the etching, does not react with CF_4 but can act in two opposite ways:

- O atoms react with unsaturated halocarbon species, eliminating polymerization and enhancing the generation of F atoms in accord with the etchant unsaturated species model; in this case, the etching is enhanced by the addition of oxygen
- When sufficient oxygen is present in the plasma, O_2 or O can occupy film adsorption sites on the processed wafer and thus inhibit etching [31].

Gases such as NF_3 and ClF_3 [32, 33] are also interesting for etching of silicon because they supply fluorine but do not contain atoms that can form residues. Therefore, high concentrations of atomic fluorine are obtained in these gases without the addition of oxygen, which can attack resist materials used for etching masks [3].

Plasmas that produce chlorine and bromine atoms are used in RIE for anisotropic etching of silicon, because these gases cannot etch silicon at room temperature without the assistance of ion bombardment. Commonly used chlorine-containing gases are Cl_2, CCl_4, CF_2Cl_2, and CF_3 as well as mixtures such as $Cl_2 + C_2F_6$, $Cl_2 + CCl_4$, $HCl + CCl_4$, and $C_2F_6 + CF_3Cl$. High etching rates (500–6300 Å/min for undoped and doped poly-Si) and high selectivities ($Si{:}SiO_2 \approx 10$–$50{:}1$) have been reached with these gases. The active etchants in chlorine or bromine plasmas are likely to be Cl and Br atoms, and ion bombardment plays a significant role in achieving high etching rates and control of anisotropy. The high degree of anisotropy that is achieved indicates that ion bombardment dominates the etching mechanism by enhancing either the reaction with the chemisorbed $SiCl_2$ through improved penetration of Cl into the Si surface or by enhancing the volatility of the etching product. These conclusions are also consistent with the fact that only a small loading effect is observed in chlorine-based etching [34], whereas fluorine-containing plasmas exhibit strong loading effects, typical of isotropic plasma etching [3].

As mentioned earlier, heavily doped ($> 10^{18}$ cm^{-3}) n-type silicon (single-crystal, or poly-Si) etch faster in Cl- and F-containing plasmas than p-type or undoped silicon [34, 35]. Because of the fast etching rate, ion bombardment is apparently not required in etching of n^+ poly-Si and isotropic etching of the highly doped n^+ poly-Si occurs often in chlorine plasmas.

8.4.1.2 Silicon Dioxide and Silicon Nitride

Silicon dioxide and silicon nitride can also be etched by fluorine atoms; however, etching rates are low without the assistance of particle bombardment due to the strength of the Si-O bond. Directional etching can be achieved in fluorine-containing gases. The etching rate of silicon dioxide, $R_F(SiO_2)$, by fluorine atoms is given in Å/min by [30]:

$$R_F(SiO_2) = 6.14 \times 10^{-13} n_F T^{1/2} \exp\left(-0.163 \frac{eV}{kT}\right) \quad (8.16)$$

The gases previously described for silicon etching are also suitable for etching silicon oxide and nitride. Plasma conditions that provide selective etching are important when the etching of these materials is performed in the presence of silicon. To achieve selective oxide and nitride etching, gases that create fluorine-deficient environment are added to the fluorine plasmas. Additives that act as efficient scavengers of fluorine are H_2, C_2H_4, and CH_4. Alternatively, molecules that contain F, C, and H, such as CHF_3, can be used.

To obtain etching selectivity of oxide and nitride over silicon, the chemistry of the plasma has to be adjusted close to the limit between etching and polymerization as discussed earlier in Sec. 8.3.3 (see Fig. 8-8). The correct gas mixture therefore has to be determined often empirically. For example, in CF_4 plasma, the selectivity of silicon dioxide over silicon is practically inexistent, (SiO_2:Si ≈ 1.3). However, if H_2 is added to obtain a CF_4-H_2 plasma, fluorine atoms are scavenged by hydrogen to form HF, the C/F ratio of unsaturated species in the gas phase is increased, and polymerization is promoted. Oxygen released from SiO_2 during etching reacts with the carbonaceous polymer and forms volatile products such as CO, CO_2, and COF_2, and etching of the oxide continues undisturbed. On silicon, no mechanism other than sputtering is active in the CF_4-H_2 plasma for carbon removal. Therefore the deposited polymer film inhibits etching of Si even at low levels of hydrogen additions. At 40% H_2, a selectivity of SiO_2:Si ≈ 40 is achieved under specific conditions of pressure, power, and so on [36].

8.4.1.3 Group III–V Materials

Group III–V semiconductor compounds include GaAs, GaP, InP, and GaAlAs, which form the basis for many new electronic applications, particularly high-speed integrated circuits, microwave, and optoelectronic devices. The applications of the III–V compounds involve binary, ternary, and even quaternary alloys. Therefore, the etchants must be capable of causing selective etching when stoichiometric changes are made to obtain specific device characteristics.

Up to the present, primarily the binary compounds have been studied, with special attention being given to the volatility of the reaction products. The group III–V compounds are particularly difficult to etch in a plasma because the group V elements form volatile halides, while halides of group III elements, particularly the fluorides, tend to be nonvolatile. As a result, fluorine-containing plasmas are generally not practical for etching the group III–V materials, unless high ion bombardment energies are used. Hydrogen discharges have been used to etch group III–V materials [37] at temperatures above 100 °C, giving low selectivities of GaAs oxide to GaAs (~ 2). GaAs etching displayed undercut profiles [3].

Most studies have used chlorine-containing plasmas at elevated temperatures to take advantage of the increase, though limited, of the volatility of the group III chlorides with increasing temperature.

Plasmas of silicon tetrachloride ($SiCl_4$) and sulphur hexafluoride (SF_6) have been used to etch GaAs and GaAlAs. Selective etching of GaAs was obtained by increasing the concentration of SF_6 in the gas mixture. Optical emission spectroscopy of these plasmas indicated an increase in concentration of Cl atoms with increasing concentration of SF_6 [38]. Selectivity is obtained because of the formation of an AlF_3 passivating layer over GaAlAs when SF_6 is added to $SiCl_4$ in the plasma. This passivating layer cannot form on GaAs. Etching selectivity increased with pressure in the range 15–100 mtorr. DC biasing of the substrate did not affect the etching rate of GaAs, while enhancing the etching rate of GaAlAs by one order of magnitude when the bias changed from -100 V to -200 V [38].

8.4.2 Metals

Metals such as aluminum, tungsten, molybdenum, and titanium can be etched easily in a plasma. Other metals, such as copper, silver, iron, nickel, and cobalt, are very difficult to be dry etched. Aluminum, copper, tungsten, and titanium are among the metals used in VLSI fabrication and have to be etched to delineate the circuits. Aluminum is widely used as interconnect layer for integrated circuits. Copper is added to aluminum to increase its resistance to electromigration. In addition to aluminum, the refractory metals and their silicides have been the subject of the most extensive efforts in metal etching.

Etching of metals requires as a first step the removal of their native oxides. Because of the high reactivity of most metals with oxygen and water vapor, plasma etching of metals may require more attention to reactor design and process details than necessary with other materials. Unless the metal forms a volatile oxyhalide or an unstable oxide, water vapor and oxygen must be excluded from or scavenged in the plasma reactor. Since the metal-oxygen bond may be extremely strong, ion bombardment is required to assist native oxide removal. This is done under high-energy ion bombardment. For more efficient effect of ion bombardment, heavier chlorocarbon or fluorocarbon gases, which are also able to reduce native oxides chemically, are used instead of halogens [3].

Halocarbon vapors are particularly susceptible to polymerization, which causes formation of residue that can interface with etching processes.

The plasma etching characteristics of Al have been extensively studied. Aluminum and chromium do not form volatile fluorides: therefore these metals are etched with chlorine-containing gases. Aluminum etches easily in chlorine because $AlCl_3$ is highly volatile at room temperature. Chlorine-containing gases (CCl_4, BCl_3, $SiCl_4$, and $CHCl_3$) have therefore been the preferred etchant for aluminum [39, 40]. A few studies have also investigated the use of brominated gases (HBr, Br_2, and BBr_3) and found results similar to those obtained in chlorinated gases. Aluminum cannot be etched with fluorine-containing gases because the aluminum fluoride is not sufficiently volatile. If aluminum fluoride is formed at a surface, further reaction is blocked and etching terminates [11].

An initiation period, or lag time, exists at the start of aluminum etching due to the scavenging or removal of oxygen and water vapor residual in the reactor and etching of the thin native aluminum oxide layer always present on aluminum surface [5]. To keep the amount of oxygen and water vapor at a minimum in the reactor, a load-lock can be used in such a way that the chamber is not exposed to air between etching runs. The etching of the native aluminum oxide can be enhanced by the ion bombardment of the surface and by supplying chemical species capable of reacting directly with the oxide such as CCl_x, BCl_x, and $SiCl_x$.

After the removal of the native aluminum oxide layer, molecular chlorine (Cl_2) can etch pure clean aluminum even without a plasma. Cl_2 rather than Cl appears to be the primary etchant species for aluminum in a glow discharge. Aluminum films often contain silicon, and because $SiCl_4$ is volatile at room temperature, aluminum-silicon films can also be etched in chlorine-containing discharges.

After plasma etching, aluminum films may often corrode upon exposure to atmospheric conditions. The corrosion is a result of the hydrolysis of chlorine-containing residues (mostly $AlCl_3$) remaining on the film side walls, on the substrate, or in the photoresist. Contamination with carbon and radiation damage caused by particle bombardment may enhance corrosion susceptibility. When copper is present in aluminum films, accelerated postetch corrosion is observed. This phenomenon occurs for two reasons: (1) hygroscopic $CuCl_2$ is probably left in contact with aluminum film because of low vapor pressure, and (2) the grain boundaries have a cathodic potential relative to the aluminum grains because most of the copper is present in the grain boundaries as $CuAl_2$.

The postetch corrosion can be prevented by restoring the passivating native aluminum oxide film. This can be done through low-temperature thermal oxidation in dry oxygen. Another method of preventing postetch corrosion is to expose the aluminum film to fluorocarbon plasmas such as CF_4 or CHF_3 [5]. This treatment converts the chloride residues into nonhygroscopic fluorides and deposits a fluorocarbon polymer film onto the Al surface so that Al can be exposed to ambient conditions without immediate corrosion.

Copper does not form volatile chlorides or other halides, and therefore its removal during aluminum etching is difficult. Two methods can be used to promote copper chloride desorption [5]:

- Increase the substrate temperature as high as allowed by the used photoresist material
- Enhance the ion bombardment so that significant sputtering or surface heating is achieved.

The fluorides and chlorides of the transition metals (molybdenum, tungsten) and their silicides are volatile in the presence of ion bombardment. Transition metals can be etched with fluorine- or chlorine-containing gases and mixtures of these gases, such as $SF_6 + Cl_2$. Oxygen can be added to enhance the etching rate of the transition metals because some oxyfluorides (e.g., those of molybdenum and tungsten) have reasonable vapor pressure.

Gold can be etched effectively with $C_2Cl_2F_4$ or $CClF_3$. Chromium is etched easily in plasmas containing chlorine and oxygen because of the high volatility of the oxychloride (CrO_2Cl_2). Titanium can be etched in fluorine-, chlorine-, or bromine-containing gases.

8.4.3 Polymers

Cold plasmas are also employed for etching patterns in polymers during fabrication of integrated electronic devices. The use of cold plasmas for etching of polymers has been described in Sec. 6.3.2, in the context of resist stripping. The possibility of using the cold plasmas, especially in the RIE mode, for anisotropic etching has extended the applications of polymer etching in various fields of submicron technology, for formation of high-resolution resist patterns with steep profiles. Plasma etching of polymers is also used to define openings in polyimide for electrical contacts between metallization layers [6].

Pattern transfer in fabrication of integrated circuits is performed by preparing suitable masks using photolithography of photoresists. Single-layer light-sensitive photoresists are photographically exposed to an image of the pattern and then developed to create the mask. In recent years lithography has been expanded to multilevel (or multilayer) bilevel and trilevel lithography schemes used for definition of high-resolution microelectronic patterns [4, 41].

With the assistance of cold plasma processing, multilayer lithography enables the achievement of high-resolution pattern transfer with simultaneous planarization of the underlying surface topography. Of particular interest are processes that employ silicon-containing, or silylated, polymer resists as the imaging layer above a hydrocarbon-based planarizing layer [42]. Information on the formulation and lithographic processing of these resists can be found in reviews such as [4, 43].

The multilevel resist is first processed by regular photolithography to replicate the pattern on the topmost layer. Next, when exposed to the plasma, the silicon near the surface of the resist is converted to a silicon oxide layer, which is resistant to the etching plasma species and thus serves as an in-situ mask for pattern transfer into the underlying planarizing layer. The multilayer pattern transfer requires etch selectivity and anisotropy between the silicon-containing and hydrocarbon-based materials. The high degree of anisotropy in multilevel etching is achieved using reactive ion etching, sometimes with magnetic confinement, ECR plasmas, or reactive ion beam etching (RIBE) [4].

Oxygen plasmas used for resist stripping are also used with photolithography for etching of patterns in polymers. The reactive species responsible for the polymer etching are oxygen atoms and atomic and molecular ions (O^+, O^{2+}) produced in the plasma. The active species responsible for the initiation of etching is the oxygen atom; however, the presence of only oxygen atoms is not sufficient for etching unless heat, ion bombardment, or fluorine atoms are also present [4]. In a microwave plasma, the etching rate of polyimide was found to be directly proportional to ion density, both with and without RF biasing of the substrate. While the main etching species of the polyimide are excited oxygen atoms, activation of polyimide surface by ion bombardment is an essential first step and was found to be the rate determining process [4].

In an RIE plasma process, the flux of highly oriented energetic ions increases the polymer etching rate and also defines the degree of anisotropy. Isotropic etching of resist can be suppressed by cooling the processed substrates, because isotropic etching is a thermally activated process and its rate increases with increasing temperature [44]. On the other hand, the shapes of the resulting profiles depend strongly on the relative fluxes of ions, molecules, and radicals. Energetic ions can sputter the polymer surface and increase the surface temperature. Additionally, ion bombardment leads to dissociation of adsorbed molecules and desorption of reaction products, as well as to damage of near-surface regions of the polymers being etched. It is widely believed that the dominant etching mechanism in submicron processes are chemically enhanced physical sputtering, physically enhanced chemical etching, or surface damage induced etching [13].

A model developed by Hartney et al. describes the etching of polymers with oxygen plasmas through a three-step mechanism [4]:

- Oxygen atoms remove the hydrogen from the surface layers leaving activated polymer fragments on the surface.
- The activated surface further reacts with molecular oxygen forming precursors of volatile reaction products.
- Finally, ion bombardment removes the volatile products and completes the etching process.

This model is in good agreement with the observed dependency of the etching rates on the neutral flux. However, the formation of passivating layers, such as in sidewall passivation, is not yet fully understood [4].

In oxygen plasmas, a direct relationship has been observed between etching rates of polymers and concentration of atomic oxygen in the plasma. Etching rates can therefore be increased in oxygen plasmas by increasing the concentration of the oxygen atoms. This can be achieved by increasing the electron density and temperature. Addition of helium to an oxygen plasma increases the electron temperature and addition of nitrogen has been found to increase the concentration of oxygen atoms [6].

Addition of even small concentrations of halogen atoms dramatically increases the etching rate of polymers in an oxygen plasma. For example, CF_4 added to oxygen plasmas increases the concentration of oxygen atoms [6]. Fluorine atoms can also react chemically with the polymer abstracting hydrogen and creating reactive radical sites on the surface of the polymer. These sites react easily with atomic and, possibly, even with oxygen molecules and promote the etching of the polymer. Chlorine-containing polymers have shown significantly higher etching rates compared to analogous hydrocarbon materials. Chlorine atoms broken from the polymer chain of chlorine-containing polymers were found to be more efficient than oxygen atoms in abstracting hydrogen and enhancing the etching [4].

In $O_2 + CF_4$ plasmas the highest etching rate is generally obtained at about 20–40% CF_4 and often exhibits a sharp maximum in this range [6, 45]. At low concentrations of CF_4, the increase in etching rate with CF_4 addition was observed to be more rapid than the increase in oxygen atom concentration, while at higher CF_4 concentrations the oxygen atom concentration was observed to remain relatively constant while the etching rate decreased strongly.

Ion bombardment was shown to be a key element in removing fluorinated species that passivate polymer surface during etching with $O_2 + CF_4$ plasma [45]. When samples were shielded from ion bombardment in a pure oxygen plasma, no etching occurred.

It has been found that, for RIE performed in a capacitively coupled diode reactor, the limiting factor of the etching rate of polymers is the flux of chemically active neutral or ionized particles onto the structure. RIE of polyimide with 1-KeV O^+ ions had an etch yield six times higher than etching with 1-KeV Ar^+ ions [46]. It was also found that RIE rates of polymers decrease with increasing aspect ratio of the etched feature. This observation has been explained by the decreasing supply of neutral particles (O, O_2) into the deep structures [41].

Anisotropic etching of polymers by RIE has been explained as being a result of sidewall passivation with a carbonized layer [41]. Such a layer has been found to form in various plasma gases, such as O_2, $SF_6 + O_2$, $Ar + O_2$, and H_2. Modified sidewalls were observed to form even during RIE of polymers in a pure, chemically inert, argon plasma, indicating that ion bombardment by itself can lead to the formation of modified sidewalls. Ion bombardment of the bottom resist can lead to direct flow of neutral etch products, especially of carbon-containing particles, onto the sidewall. However, the effect of neutral plasma species on the formation of sidewalls is important. Sidewall modifications were found to decrease with increasing aspect ratios when fewer neutrals can reach into the structure, indicating a combined interaction of neutrals and ions for the formation of modified sidewalls or even synergetic reactions induced by neutral and ion fluxes on the sidewalls [41]. The release of carbon molecules from parts of the reactor, such as electrodes, walls, extraction grids, and so on, can also contribute to the neutral flux onto the sidewall.

The etching mechanism of organosilicon polymers in an oxygen plasma has been explained by Watanabe and Ohnishi through a steady-state model [47]. In high-pressure oxygen plasmas, when the polymers do not undergo ion bombardment, organosilicon polymers are oxidized into silicon dioxide. However, when during processing in an oxygen plasma a bias voltage is applied to the substrate (i.e., when the polymers are bombarded by high-energy ions), only a thin steady-state layer of silicon dioxide is formed on top of the organosilicon polymer. The steady state is reached between physical sputtering and chemical reactivity, when silicon removal by sputtering is balanced by regeneration through oxidation of the organosilicon polymer, in such a way that the thickness of the silicon oxide remains constant. A direct correlation exists between steady-state etching rates and the atomic silicon concentration in the silylated resist. It has been shown that the etching rate of silicon-containing polymers in oxygen plasmas is inversely proportional to the amount of silicon in the polymer [48].

The steady-state etching of the silylated polymers is not always reached. Studies of plasma etching of silylated areas in a distributed ECR reactor have shown that steady-state etching was not achieved in the specific etching conditions. Instead, the etching rate was found to increase with time [44]. In the high-density ECR plasma a good silicon dioxide barrier layer was observed to form already at room temperature, while in a low-density RIE plasma the formation of a good barrier layer was obtained only at elevated temperature. The steady-state etching rate in RIE decreased with increasing wafer temperature.

The etching rate of polymers is temperature dependent and has been found to increase with increasing temperature. The temperature dependence of the etching rate of polymers can be described by an activation energy E_a through the Arrhenius equation [6]:

$$R = A \exp\left(- \frac{E_a}{k\mathrm{T}} \right) \tag{8.17}$$

where R = etching rate
A = pre-exponential factor
E_a = activation energy

Wafer temperature is therefore one of the most important parameters to be controlled during polymer etching [44].

An important technological problem that has yet to be solved is the relative slow etching rate obtained in plasmas used to obtain etch anisotropy, especially with silylated polymers. If the RF power is increased in RIE of silylated polymers, the self-bias and ion energies increase. As a result, the etching rates of both the silicon-containing resist and planarizing material increase, but at the same time, etching selectivity generally decreases.

8.5 ETCHING INDUCED DAMAGE

As already discussed in Sec. 3.5.4 and mentioned in Chapters 6 and 7, one has to be aware that exposure to plasmas can induce damage to the surface layers of sensitive materials. The main cause of this damage is the bombardment of the surface with ions and energetic neutral plasma particles, although electron and UV bombardment might also contribute sometime to the damage. While any surface in contact with a plasma is exposed to bombardment with energetic particles, the problem is especially significant in reactive ion etching. In RIE the processed substrates are negatively biased to several hundred volts, the pressure is relatively low, and as a result the substrates are bombarded with high-energy particles. In addition to the substrate damage induced by bombardment with energetic particles, chemical modifications of the etched surfaces may also cause deterioration of sensitive devices.

Special attention has to be given to the analytical tools chosen for the determination of the damage. Analytical techniques that are sensitive to plasma induced modifications of the surface might not reveal information about damage to the bulk, and vice versa. Following are a few more examples of substrate damage caused by cold plasma processing and the means to determine it.

Substrate damage caused in silicon by etching in a magnetically confined triode reactor, in which the plasma was excited at 13.56 MHz and the substrate was biased with an electric field of 100 kHz, was studied by *thermal wave (TW) measurements* [49]. Surface damage or contamination were correlated to the intensity of the TW signal.

The TW signal was found to increase with increasing RF power at 100 kHz used for substrate biasing, indicating that the damage increased with the increased kinetic energy of the bombarding ions. The substrate bias was found to increase linearly with the power; however, the TW signal, which reflected the induced substrate damage, reached a maximum and leveled off at higher powers. In the discussed arrangement, the substrate degradation was found to decrease with increasing plasma exciting RF power at 13.56 MHz. This was explained by an increased rate of generation of excited species and a parallel decrease of their kinetic energy with increasing exciting power in the magnetically confined reactor [49]. The results indicated that substrate damage was caused by implantation of highly energetic reactive species and that absorbed etchants and/or nondesorbed etch products played a minor role.

Raman spectroscopy studies of GaAs subjected to RIE in argon have shown that the depletion depth in GaAs increased with increasing substrate bias during etching. A depletion depth of 273 Å was observed in a sample etched at − 350 V versus 176 Å in the unetched sample [50]. However, low-temperature cathodoluminiscence measurements of the same sample showed the damage to extend to a depth of 1000 Å, a depth much larger than the penetration range of 350 eV ions. The deeper damage was explained assuming significant room temperature diffusion of the generated defects. It was suggested that some of the damage could be reduced by performing rapid etching, which minimizes the total integrated ion doze [50]. The author also expected that direct bombardment of the device material, even at energies of ~ 100 eV, could produce optical damage extending deep into the material, a problem most critical for low-dimensional structures with relatively large surface-to-volume ratios [50].

Damage can be induced also in polymers when processed in cold plasmas. The polymers are susceptible to damage by electron, ion, and excited atom bombardment and to UV radiation. The surface bombardment causes breakage of polymer bonds and the resulting polymer fragments can crosslink. This crosslinking can result in the formation of a brittle polymer layer, a few hundred angstroms thick. Polymer films are often under strain, and the brittle polymer layer can act as a crack initiator [51]. XPS analysis of plasma treated polyimide surfaces has shown a significant increase in oxygen content, the formation of polycarbonate or peroxy groups, and an increase in carbonyl and ether groups as a result of the treatment. Contact angle measurement showed a strong decrease in water contact angle, indicating changes in the surface energy of the polymer.

8.6 QUESTIONS

1. What types of material damage can be induced by plasma etching, and what processes are responsible for these damages?
2. a. Define etching anisotropy and discuss how it can it be achieved.
   b. What effects of plasma chemistry are assisting in achieving etching anisotropy?
3. Which plasma method would you choose to achieve etching anisotropy with minimal damage to the etched material and substrate?
4. How can plasma etching be affected by surface contamination of the etched sample?
5. What is etching selectivity and how can it be achieved in plasma etching? Which plasma method is most suitable for achieving etching selectivity?
6. Search the literature for most recent studies of plasma etching of silicon. Critically discuss the findings in three selected studies.
7. Repeat question 5 for silicon dioxide and silicon nitride.
8. Repeat question 5 for oxides other than silicon oxide.
9. Repeat question 5 for three different metals and alloys.

8.7 REFERENCES

[1] Einspruch, N. G., and D. M. Brown, eds., *VLSI Electronics Microstructure Science*. New York: Academic Press, 1984.

[2] Flamm, D. L., In *Plasma Etching*, eds. D. M. Manos and D. L. Flamm, p. 91. New York: Academic Press, 1989.

[3] Hess, D. W., and D. B. Graves, In *Microelectronics Processing*, eds. Dennis W. Hess and Klavs F. Jensen, p. 377. Washington, DC: American Chemical Society, 1989.

[4] Hartney, M. A., D. W. Hess, and D. S. Soane, *J. Vac. Sci. Technol.*, B7: 1 (1989).

[5] Oehrlein, G., In *Handbook of Plasma Processing Technology*, eds. Stephen M. Rossnagel, Jerome J. Cuomo, and William D. Westwood, p. 196. Park Ridge, NJ: Noyes Publications, 1990.

[6] Egitto, F. D., V. Vukanovic, and G. N. Taylor, In *Deposition, Treatment and Etching of Polymers*, ed. R. d'Agostino, p. 321. New York: Academic Press, 1990.

[7] Flamm, D. L., and V. M. Donnelly: *Plasma Chem. Plasma Process.*, 1: 317 (1981).

[8] Irving, S. M., K. E. Lemons, and G. E. Bobos, *U.S. Patent*, 3 615 956 (1969).

[9] Flamm, D. L., *Plasma Chem. Plasma Process.*, 1: 37 (1981).

[10] Mogab, C. J., *J. Electrochem. Soc.*, 124: 1262 (1977).

[11] Graves, D. B., *AIChE J.*, 35: 1 (1989).

[12] Coburn, J. W., In *Diamond and Diamond-like Films and Coatings, NATO-ASI Series B: Physics*, eds. R. E. Clausing, L. L. Horton, J. C. Angus, and P. Koidl, Vol. 266, p. 73. New York: Plenum Publishing, 1991.

[13] Flamm, D. L., and G. K. Herb, In *Plasma Etching*, eds. D. M. Manos and D. L. Flamm, p. 1. New York: Academic Press, 1989.

[14] Mauer, J. L., J. S. Logan, L. B. Zielinski, and G. C. Swartz, *J. Vac. Sci. Technol.*, 15: 1734 (1978).

[15] Tu, T. J., T. J. Chuang, and H. F. Winters, *Phys. Rev.*, B23: 823 (1981).

[16] Kay, E., J. Coburn, and A. Dilks, In *Topics in Current Chemistry, Plasma Chemistry III*, eds. S. Veprek, and M. Venugopalan, p. 1. Berlin: Springer-Verlag, 1980.

[17] Swartz, G. C., and P. M. Schaible, *J. Vac. Sci. Technol.*, 16: 410 (1979).

[18] Ogryzlo, E. A., D. E. Ibbotson, D. L. Flamm, and J. A. Mucha, *J. Appl. Phys.*, 67: 3115 (1990).

[19] Herb, G. K., D. J. Rieger, and K. Shields, *Solid State Technol.*, 30: 109 (1987).

[20] Oehrlein, G. S., J. F. Rembetzki, and E. H. Payne, *J. Vac. Sci. Technol.*, B8: 1199 (1990).

[21] Coburn, J. W., and H. F. Winters, *J. Vac. Sci. Technol.*, 16: 371 (1979).

[22] Oehrlein, G. S., and J. F. Rembetzki, *Plasma-based Dry Etching Techniques in Silicon Technology*, IBM RC 16798 (1991).

[23] Oehrlein, G. S., *Mat. Sci. Eng.*, B4: 441 (1989).

[24] Nguyen, S. V., G. Christman, D. Dobuzinsky, and D. Harmon, *Solid State Technol.*, 33(10): 73 (1990).

[25] Singer, P. H., *Semicond. Int.*, 46 (July 1991).

[26] Cotler, T. J., and M. E. Elta, *IEEE Circuits and Devices*, 6: 38 (1990).

[27] Winters, H. F., and J. W. Coburn, *J. Vac. Sci. Technol.*, B3: 1376 (1985).

[28] Yarmoff, J. A., and F. R. McFeely, *Surf. Sci.*, 184: 389 (1987).

[29] McFeely, F. R., J. F. Morar, and F. J. Himpsel, *Surf. Sci.*, 165: 277 (1986).

[30] Flamm, D. L., V. M. Donnelly, and J. I. Mucha, *J. Appl. Phys.*, 52: 3633 (1981).

[31] Mogab, C. J., A. C. Adams, and D. L. Flamm, *J. Appl. Phys.*, 49: 3796 (1978).

[32] Flamm, D. L., D. N. Wang, and D. J. Maydan, *J. Electrochem. Soc.*, 129: 2755 (1982).

[33] Picard, A., and G. Turban, *Plasma Chem. Plasma Process.*, 5: 333 (1985).

[34] Engelhard, M., and S. J. Schwarzl, *J. Electrochem. Soc.*, 134: 1985 (1987).

[35] Lee, Y. H., and M. Chen, *J. Vac. Sci. Technol.*, B4: 468 (1986).

[36] Ephrath, L. M., *J. Electrochem. Soc.*, 126: 1419 (1979).

[37] Chang, R. H. P., C. C. Chang, and S. J. Durack, *J. Vac. Sci. Technol.*, 20: 45 (1982).

[38] Salimian, S., and C. B. Cooper III, *J. Vac. Sci. Technol.*, B6: 1641 (1988).

[39] Herb, G. K., R. A. Porter, P. D. Cruzan, K. Agraz-Guerena, and B. R. Soller, *Electrochem. Soc. Ext. Abstr.*, 81-2: 710 (1981).

[40] Purdes, A. J., *J. Vac. Sci. Technol.*, A1: 712 (1983).

[41] Pilz, W., J. Janes, K. P. Muller, and J. Pelka, *Proc. SPIE*, 1392: 84 (1990).

[42] Hatzakis, M., J. Paraszczak, and J. Shaw, In *Proc. Microcircuit Engineering '81*, ed. A. Oosenberg, p. 386. Lausanne: Swiss Federal Institute of Technology, 1981.

[43] Onishi, Y., M. Suzuki, K. Saigo, and Y. Saotome, *Proc. SPIE*, 539: 62 (1985).

[44] Dijkstra, J., *Proc. SPIE*, 1466: 592 (1991).

[45] Egitto, F. D., F. Emmi, R. S. Horwath, and V. Vukanovic, *J. Vac. Sci. Technol.*, B3: 893 (1985).

[46] Vanderlinde, W. E., and A. L. Ruoff, *J. Vac. Sci. Technol.*, B6: 1621 (1988).

[47] Watanabe, F., and F. Ohnishi, *J. Vac. Sci. Technol.*, B4: 422 (1986).

[48] Paraszczak, J., E. Babich, J. Heidenreich, R. McGouey, L. Ferreiro, N. Chou, and M. Hatzakis, *Proc. SPIE*, 920: 242 (1988).

[49] Engelhardt, M., *Proc. SPIE*, 1392: 38 (1990).

[50] Hu, E. L., *Proc. SPIE*, 1392: 512 (1990).

[51] Mukkavilli, S., R. W. Pasco, M. S. Farooq, and M. J. Griffin, In *Proc. 40th Electronic Components & Technology Conf.*, May 20-23, 1990, Las Vegas, NV (IEEE, 1990), p. 737.

Index

A

B

C

D

E

F

Q

R

U

V

W

X

Y

Z

About the Author

Dr. Alfred Grill received his Ph.D. in physics from the Hebrew University of Jerusalem in Israel in 1973. For fifteen years he was a professor at the Materials Engineering Department at the Ben-Gurion University in Israel, where he taught and performed studies of cold plasmas, plasma diagnostics, and applications of cold plasmas for deposition of films and surface treatments.

During that time he was also a visiting scientist at Max-Planck Institute for Steel Research in Dusseldorf, Germany and at NASA-Lewis Research Center in Cleveland, Ohio. He served as a member of the international organizing committee of the 10th International Symposium on Plasma Chemistry, Bochum, Germany, in 1991, and the International Applied Diamond Conferences, 1991-1995. Dr. Grill is also a member of the editorial board of the *Diamond Films and Technology* journal.

In 1989, Dr. Grill joined IBM Research Division, Thomas J. Watson Research Center, where he is a research staff member in the Systems Technology and Science department. His focus is in plasma enhanced CVD growth and tribological properties of thin hard films.

Dr. Grill has seventeen years of experience in the plasma-assisted techniques for deposition of films and surface treatment of materials. He has more than 100 publications in peer reviewed journals, as well as numerous patents and patent applications. Dr. Grill is also the recipient of the American Society for Metals' Marcus A. Grossman Young Author Award, IBM Invention Achievement Awards, and an IBM Outstanding Innovation Award.

Printed and bound by CPI Group (UK) Ltd, Croydon, CR0 4YY

16/09/2026

14963445-0001